Collins Edexcel [barcode] on

Grade Booster

Maths
Higher

Rosie Benton and Jenny Hughes

Contents

1 Introduction

About this Book

This book has been designed to support your preparation for the higher tier Edexcel GCSE mathematics exam papers. There are sections with advice and ideas about how to revise, what you need to know and ways to learn it. Other chapters contain examples, split by topic, to give you an idea of what higher tier exam questions might look like and what good answers to those questions would be.

The non-calculator symbol 🔢 appears beside questions that you are expected to be able to answer without a calculator. You should remember that a similar question could still appear on a calculator paper. When attempting these questions, try them without a calculator. It can be easy to make calculation errors, so it is important to practise even if it seems a relatively easy step in the question.

This book aims to help you to apply the knowledge you already have in your 'mathematical toolbox'. It focuses on how to implement the skills you already have in a variety of contexts.

Students often lose marks through misconceptions, feeling lost when a question is unfamiliar or making silly mistakes. This book aims to highlight ways to get started, build confidence, spot possible misconceptions and show ways of checking your working as you go.

There are many different ways to approach mathematical problems. It is worth having a go at each example question yourself before looking through the answer – it may be that you have an even more concise way of doing it, or a method that makes more sense to you. Different approaches are fine as long as you obey mathematical logic and show clearly what steps you have taken. Answers should be the same whatever method you use. You can work through the examples in order or dip in and out, whatever your preference. The examples have hints and tips alongside them. The symbol ✓ helps to show where individual marks would be awarded in the workings and answers.

Questions on the exam papers can be expected to be worth up to 6 marks. However, to get the most out of the examples, some questions in this book are allocated more than 6 marks. Any aspect of these examples could represent part of an exam question, but you wouldn't necessarily expect to see them all included in one single question.

At the end of each chapter, you are signposted to pages in the *Collins Edexcel Maths Higher Revision Guide* (ISBN 9780008112622) for more information on the topics covered. The same page references apply to the *Collins Edexcel Maths Higher All-in-One Revision & Practice* book (ISBN 9780008110369).

Terms in **bold** are among those defined in the Glossary at the back of the book.

The Edexcel GCSE (9–1) Course

The syllabus has been developed with the aim of increasing your ability to transfer skills between topics, school subjects and on into higher education or work. You will need to apply a mixture of mathematical skills in new situations and to solve problems. There is a focus on your ability to reason mathematically, both to draw conclusions and to consider accuracy within the work you are doing. The ability to communicate clearly and concisely is also important. You are expected to have a good grasp of key conventions and mathematical notation so that your work can be understood universally. Understanding is key and, as there are generally many ways to answer the questions you will be asked, your ability to show your understanding and explain your method becomes all the more important.

The Exams

- You will sit three exam papers. The first paper is a non-calculator paper, testing your mental and written methods alongside particular topic knowledge. The other two are calculator papers, allowing more focus on the context and strategies, rather than the 'number crunching'. All topics studied can appear on any of the three papers.
- Each paper has 80 marks and lasts 1 hour and 30 minutes. When practising exam questions, this works out at roughly 1 mark a minute.
- There are two tiers of entry – higher (grades 4 to 9) and foundation (grades 1 to 5). This book supports the higher tier specification. The three papers that you will sit all have to be from the same tier and be taken in the same exam series. That means you will be sitting three higher tier papers, either in the summer or possibly (if you are resitting and over 16 years of age) in November.

Higher Tier Topic Overview

The topics will be assessed based on the following weightings:

Topic	Weighting
Algebra (see Chapter 4)	27–33%
Ratio, Proportion and Rates of Change (see Chapter 5)	17–23%
Geometry and Measures (see Chapter 6)	17–23%
Probability and Statistics (see Chapters 7 and 8)	12–18%
Number (see Chapter 3)	12–18%

Assessment Objectives Overview

The style of questions are categorised in the following way:

Assessment Objective and Weighting	Requirements
AO1 Use and apply standard techniques 40%	• Accurately recall facts, terminology and definitions (<10%) • Use and interpret notation correctly • Accurately carry out routine procedures • Accurately carry out set tasks requiring multi-step solutions
AO2 Reason, interpret and communicate mathematically 30%	• Make deductions and inferences to draw conclusions from mathematical information • Construct chains of reasoning to achieve a given result • Interpret and communicate information accurately • Present arguments and proofs • Assess the validity of an argument and critically evaluate a given way of presenting information
AO3 Solve problems within mathematics and in other contexts 30%	• Translate problems in mathematical or non-mathematical contexts into a process or a series of mathematical processes • Make and use connections between different parts of mathematics • Interpret results in the context of the given problem • Evaluate methods used and results obtained • Evaluate solutions to identify how they may have been affected by assumptions made

The different types of question give you the opportunity to show that you can:
• remember key facts and use them to solve a problem
• solve problems by making use of your mathematical knowledge in less familiar situations.

This book mainly focuses on the AO2 and AO3 styles of questions.

See Chapter 9 for examples, hints, tips and methods for dealing with the different types of questions.

> *"Luck is not chance –*
> *It's Toil –*
> *Fortune's expensive smile*
> *Is earned –"*
>
> — Emily Dickinson

Mathematics is a skills-based subject so the most useful way to prepare is to practise, practise, practise. There are some knowledge-based elements, formulae and facts you need to remember. More detail about what you 'need to know' can be found in Chapter 10 (see page 136).

Planning Your Revision

However much time you have left before your exam, there is always a lot that you can do. Think carefully about what you are doing and how you are doing it.

Long-term Preparation

The key with maths is making sure you understand everything as it is taught, and then revisiting topics regularly, especially ones you find challenging. Think about building in a regular slot to look over past topics. Make notes and refine those notes as you go. Doing small amounts of regular work means that there is less need for cramming, and stress, when it comes to an exam.

Shorter-term Preparation

If you have relatively limited time left, make sure you plan how to use that time most efficiently (do not spend too long planning though). It can be tempting to work on a topic that makes you feel good because you can do it. However, in terms of adding marks to your exam performance, it is likely to mean a lot of time for perhaps only a few extra marks. Instead, pick out topics that you know you struggle with, find different ways to work on them and feel good when you conquer something that you started out dreading.

Be Organised

Whether you are someone who is naturally organised or chaotic, it is worth putting in place a structure to help you keep on top of what you are doing, for example:
- Lists of topics that you can tick off for showing you have learned/reviewed/are confident in. Colour coding can work well.
- A record of which past papers you have already done, how you did and what specific things you need to improve.

Your teacher might be able to provide you with a template for these or you might need to do some research yourself.

Create Time and Space

In order to focus on your work, give yourself a positive environment to work in. A quiet space where you are unlikely to be interrupted is a good start. Turn off your phone and the television. You might think you can multi-task, but give your brain a chance to focus. Make it comfortable – working on a desk gives you much better posture and can help you to concentrate.

Be Kind to Yourself

In the run-up to exams it can be easy to intend to dedicate every last minute to revision but then get distracted and put it off. This cycle can make you feel guilty and overwhelmed. To help you work hard and remain focused, plan to take breaks and take them guilt-free (as long as there aren't too many and they aren't too long!). Your brain needs some downtime to process the information you have been revising.

A good revision plan allows some overflow sessions. These can be used to do a bit extra on something you run out of time for, without disrupting the rest of the plan. If you had time for everything, you can then have a bonus session off. Try to be realistic about what you can achieve in a one-hour or two-hour session.

Eat, Drink, Sleep and Exercise

Give your body and your brain what it needs to function at the top level.

Enjoy It

This may sound ridiculous, or completely obvious to you, but problem-solving and logic puzzles can be really rewarding and enjoyable, as shown by the popularity of number puzzles like Sudoku. Sometimes it is hard to enjoy the process because you are so focused on the final goal. Enjoy each question and activity for what it is. Don't panic! When you get stuck, it is a chance for you to solve something really tricky, which is far more rewarding than doing something that comes easily to you.

Practise, Practise, Practise

Maths is a skill in applying logic and problem-solving. The best way to improve is to practise, and practise regularly. Practising maths isn't just about working through piles of past exam questions, though that should be part of it:

Learning and Revising Mathematics

Exam-style questions	It is important to be aware of how questions are likely to look in an exam so that it isn't too confusing when you open your paper.
Puzzles and activities	In lessons you are likely to have come across a mixture of activities to develop understanding of different topics. Consider using these in your own revision. You might want to make something yourself, ask a teacher or find something from a website.
What else?	Ask yourself what else a question could ask. If there was an extra part, what might it ask? If you were writing this exam paper, how might you make this question harder? This will allow you to get even more out of every question.
Spot the 'tricks'	Examiners don't set out to trick you but exam questions will include elements that are likely to trip up candidates. Spend time reading through questions and thinking about what might be the thing that would catch out candidates. Spotting what makes a question harder is a great way of planning how to get around the difficulty.
Calculators	Being familiar with your calculator is very important (whether it is your personal one or a school one). Take it to all your lessons and make sure you can use it efficiently and effectively. Always question your calculator; if your calculator suggests a door is 30 m tall, it may be worth double checking. There is a risk that you trust your calculator more than your common sense – the calculator will give you the correct answer to what you input but it is easy to make mistakes in the process. Your method is still very important so write down any calculations you do with your calculator, not just the answer.
Non-calculator questions	Don't be tempted to use a calculator when practising a non-calculator question. It is easy to decide to save time by bypassing the time-consuming process of carrying out written calculations, especially if a calculator is sitting right there. Calculation errors and silly mistakes are very common. By focusing on and practising calculations, even when they seem easy, you are more likely to spot mistakes and reduce errors. Marks will be awarded for method, so you should get used to showing clearly what you are doing.
Check your answers	Don't reach for the answer booklet too soon. Try putting in some self-checking processes and seeing if you can find any mistakes yourself.

How to Learn Things

Different people's memories work in different ways. Here are some ideas of how to help you memorise the information that you need, given that developing a photographic memory is unlikely:

Prove it/ understand it	Understanding where a fact or **formula** comes from means that, rather than just remembering the formation of letters, you know why it appears the way it does. This might also help you work out the fact or formula even if you cannot remember it. For example, many students lose marks by misremembering the laws of indices. If you know that there is a rule to answer $a^{12} \times a^5$, think about what an index means. Consider a simple case, like $a^2 \times a^3 = a \times a \times a \times a \times a = a^5$, to confirm that you need to add the powers. $a^{12} \times a^5 = a^{17}$
Make a link	Making a link might be a mnemonic to remember SOH CAH TOA (Some Old Horse Caught Another Horse Taking Oats Away) or a formula. For example, speed can be measured in miles per hour (something you should already know so do not need to learn): miles measure distance, hour is time and 'per' means divide, so Speed $= \frac{\text{Distance}}{\text{Time}}$
Write it out	To help something stick, writing it out can help as it connects with your muscle memory. In turn, this gives your brain an extra connection.
Create flashcards	Double-sided cards are easy to carry around and handy for testing yourself and your friends. They are useful for formulae but also for things like recognising key features of numbers.
Do something active	Walk the lines in an angles question, turning through the angles. Some people's brains really thrive by linking a physical movement with the fact.
Do something peculiar	The more links you can make the better. If you can connect a funny or odd event or activity with a fact that you have to learn, it gives your brain more cross-references with which to unlock the information.
Make posters	By making a poster you will consider the topic, what goes together and what similarities and differences there are. You might link visual images with an **equation** or value. All of this creates useful connections which help your brain to cross reference, store and find information when you need it.
Use your posters	Put up your posters somewhere you will see them often. Consider moving the posters around every now and then so they don't start blending into the background.
Use it	As part of your practice, you will use your store of knowledge, which in turn will help you to remember it. Think about ways to test yourself. It can be easy to think you know everything if it is written down next to you as you work and, whilst it will be filtering in slowly, there are things you can do alongside the questions to boost your uptake: • Work without any formulae and note down the ones you had to look up; use the formulae for a poster, copy them out a few times and stick them on notes by your bed. Ensure that you won't need to look them up next time. • Make a lift-the-flap formulae guide, letting you see what formulae there are available to you without giving them away. You can also add a hint flap to help build a connection.

Luck is 95% hard work and 5% good fortune... nevertheless – GOOD LUCK!

Number

Good numeracy skills underpin nearly every part of mathematics. Number takes the study of numbers a step further. Topics at GCSE look at how you are able to manipulate, interpret and express numbers in order to solve a range of problems.

3.1 Comparing and Converting Numbers

This type of question can be given as a set of numbers in different forms for you to list in a particular order, or can be given context. Your main aim is to convert them into a form that means you can compare them easily.

You can compare the numbers in various ways. Decimals will always give an easy comparison once found. Converting to decimals tends to be relatively straightforward too. If you need to do extra written methods to do the calculations, there should be plenty of space to do them. Never feel you should be able to do something in your head – especially as in exam conditions your ability to do mental maths can be put under extra pressure.

Example 3.1 📱

List the following values in ascending order:

46.5, $\frac{5}{7}$ of 63,   12.5% of 368, $\frac{937}{20}$, $45\frac{7}{25}$ *(4 marks)*

Finding values as decimals:

46.5	$\frac{5}{7}$ of 63	12.5% of 368	$\frac{937}{20}$	$45\frac{7}{25}$	
46.5	$= (63 \div 7) \times 5$ $= (9) \times 5 = \mathbf{45}$	10% of 368 = 36.8 5% of 368 = 18.4 2.5% of 368 = 9.2 12.5% of 368 $= 36.8 + 9.2 = \mathbf{46}$	$= (937 \div 2) \div 10$ $= (450 + 15 + 3.5)$ $\div 10$ $= 468.5 \div 10$ $= \mathbf{46.85}$	$= 45 + 7 \div 25$ $= 45 + 7 \times 4 \div 100$ $= 45 + 28 \div 100$ $= 45 + 0.28$ $= \mathbf{45.28}$	✓ ✓ ✓
4	Smallest (1)	3	5	2	

Values in <u>ascending</u> order:

$\frac{5}{7}$ of 63, $45\frac{7}{25}$,   12.5% of 368,   46.5, $\frac{937}{20}$ ✓

Your final answer should have the numbers in their original form. Make sure your list is in the right order.

3.2 Recurring Decimals and Fractions

To convert from a fraction to a decimal, you treat it as a division. Using a written method is a good way of doing this. You should be comfortable using the notation to express recurring decimals accurately (that is, without truncation or rounding):

- Sometimes you will get a recurring decimal with one digit that repeats, e.g. $\frac{1}{3} = 0.333333... = 0.\dot{3}$
- A recurring decimal can also have a sequence of digits that repeats, e.g. $\frac{2}{13} = 0.153846153846... = 0.\dot{1}5384\dot{6}$

The dots show where the pattern is repeated from and to.

Example 3.2

a) Express $\frac{2}{15}$ as a decimal. *(2 marks)*

$\frac{2}{15} = 2 \div 15$ ✓

$15\overline{)2.^20^50^50...}$ 0.1 3 3...

So, $\frac{2}{15} = 0.1\dot{3}$ ✓

> Make sure you set up your written method clearly. In this case, it is using short division. If you are using short division, remember that you need to add your decimal point and can continue to add zeros after it until you have found the pattern that recurs.

b) Express 0.024 as a fully simplified fraction. *(2 marks)*

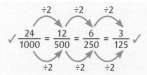

$$✓ \frac{24}{1000} = \frac{12}{500} = \frac{6}{250} = \frac{3}{125} ✓$$

> If the decimal terminates, look at the digit furthest to the right and consider its place value. In this case, the 4 is in the thousandths column, so you put 24 over 1000 and then look to simplify.

c) Prove algebraically that the recurring decimal $0.1\dot{0}\dot{6}$ can be written as $\frac{7}{66}$. *(2 marks)*

> 'Prove' means you have to start with the original situation. In this case you must start from the decimal and show you can write it in fraction form. Doing $7 \div 66$ would be showing that the fraction can be written as the decimal, i.e. the opposite way round to what the question is asking for.

	Let F =	0.106060...
①	10F =	1.060606...
②	1000F =	106.060606... ✓
②-①	990F =	105

$$F = \frac{105}{990} = \frac{21}{198} = \frac{7}{66} ✓$$

> 'F' has been used in this case but you could use any other letter to represent the number.
>
> Remember: the aim is to line up the recurring digits after the decimal point so when you subtract one equation from the other they will cancel each other out, leaving an integer on both sides.

To get the decimals to match, you first need to get the repeating pattern to start straight after the decimal point. If the number was 0.80345345345... you would first multiply by 100 to get $100F = 80.345345345...$. Then you need to do another multiplication so that one set of the repeating pattern is moved past the decimal point. In this case that would be a further 1000 (as there are three digits in the repeat) meaning your new value would be $100\,000F = 80\,345.345345345...$. If you choose to multiply by $1000F = 803.45345345...$ and then $1\,000\,000F = 803\,453.45345345...$, the decimals will still line up so you can proceed with the answer. You will just have a more complicated job when it comes to simplifying the fraction.

3.3 Standard Form

Standard form is used to express numbers that are very big or very small. It allows you to compare and calculate with these numbers without writing, counting and dealing with long strings of zeros. You need to be able to convert from and to standard form. You also need to be able to calculate with numbers in standard form, both with and without a calculator.

Example 3.3.1 🖩

The wavelength of blue light is approximately 0.00000045 m. The wavelength of red light is approximately 7×10^{-7} m.

a) Which has the greater wavelength, and by how much? Give your answer in standard form. *(2 marks)*

The wavelength of blue light = 4.5×10^{-7} ✓

The wavelength of red light = 7×10^{-7}

$7 \times 10^{-7} - 4.5 \times 10^{-7} = 2.5 \times 10^{-7}$

Red light has the greater wavelength, by 2.5×10^{-7} m. ✓

You can convert both numbers into standard form or into normal numbers to compare and calculate. Sometimes one way will be easier than the other. In this case, both numbers are 'of the order' $\times 10^{-7}$ so the subtraction was relatively easy, even in standard form.

Including a unit in your final answer is really helpful and demonstrates your understanding. It also helps you to spot where units are used and avoid mistakes in cases where units might need converting.

b) An X-ray has a wavelength that is 0.0034 times that of blue light. What is the wavelength of this X-ray? Give your answer in standard form. *(2 marks)*

$$\begin{array}{r} 34 \\ \times\ 45 \\ \hline 170 \\ 1360 \\ \hline 1530 \end{array}$$

Without a calculator this can seem fairly daunting. If you can identify what makes it hard then you can find a solution. By using powers of 10, you can make it so your calculation is dealing with integers and the powers of 10 separately.

$0.0034 = 3.4 \times 10^{-3} = 34 \times 10^{-4}$

$4.5 \times 10^{-7} = 45 \times 10^{-8}$

$4.5 \times 10^{-7} \times 3.4 \times 10^{-3} = 34 \times 45 \times 10^{-4} \times 10^{-8}$ ✓

$= 1530 \times 10^{-12}$ ←

$= 1.53 \times 10^{-9}\,\text{m}$ ✓

This is not yet in standard form but students often lose the last mark by not realising that.

Example 3.3.2

The following table gives the masses of four stars in the Milky Way. There are estimated to be 80 000 000 000 stars in the Milky Way.

The Sun	Sirius	Betelgeuse	Proxima Centauri
$1.989 \times 10^{30}\,\text{kg}$	$4.018 \times 10^{30}\,\text{kg}$	$1.531 \times 10^{31}\,\text{kg}$	$2.446 \times 10^{29}\,\text{kg}$

Using the information above, and making clear any assumptions, approximate the total mass of the stars in the Milky Way. *(4 marks)*

Assuming that the sample of four stars is representative of all the stars in the Milky Way:

There are different approaches you could take and there isn't a single correct final answer. The clearer you can explain what you have done, and why, the easier it is for an examiner to give you full credit.

Mean mass of a star $= \dfrac{1.989 \times 10^{30} + 4.018 \times 10^{30} + 1.531 \times 10^{31} + 2.446 \times 10^{29}}{4}$ ✓

$= 5.3904 \times 10^{30}\,\text{kg}$

Estimate for total mass
$=$ Mean mass $\times$ Number of stars ✓

Approximation isn't just guessing! Make it clear what you are doing and why.

$= 5.3904 \times 10^{30} \times 8 \times 10^{10} = 4.31232 \times 10^{41}$ ✓

An approximation for the mass of all the stars in the galaxy is $4.31 \times 10^{41}\,\text{kg}$ (3 s.f.) ✓

Here you are being asked to use a small sample to calculate a value for all the stars. You need to make an assumption to use this information, so it is an approximation as you don't know if the sample fairly represents the whole. If asked to approximate or estimate, think about the context and decide if it means using accurate data but extending the model based on assumptions, or if it means rounding numbers to do the calculation.

If using a calculator, make sure you show what you key into it, as well as the answer it gives. This means you will still get method marks even if you make a mistake keying in the calculation.

3.4 Factors and Multiples – LCM in Context

Questions looking for the lowest common multiple (LCM) will often be given in context and therefore be asking for something such as 'the least number of packets' or the first time something happens.

Example 3.4

Two buses, the 590 and the 592, do a circular route through Barton. The 590 completes a loop every 20 minutes. The 592 completes a loop every 45 minutes. Both the 590 and the 592 leave Barton at 8 am.

a) When is the next time both buses will be in Barton together? *(2 marks)*

$20 = 2 \times 2 \times 5$

$45 = 3 \times 3 \times 5$

$LCM = 2 \times 2 \times 5 \times 3 \times 3 = 180$ ✓

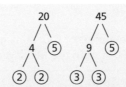

The buses will meet up in Barton every 180 minutes, or 3 hours. So the first time they will be together is 11 am. ✓

There are other methods for finding the LCM. This example shows **prime factor** decomposition, then multiplying up to find the LCM.

b) One morning, the 590 is delayed in setting off by 10 minutes.
Will the buses still meet up in Barton? Justify your answer. *(2 marks)*

Times that the 590 is leaving Barton are
08:10, 08:30, 08:50, 09:10, 09:30, ...
Times that the 592 is leaving Barton are
08:00, 08:45, 09:30 ✓
The buses will meet again at 09:30, then every 3 hours through the day from then on. ✓

3.5 Factors and Multiples – HCF in Context

When finding the highest common **factor** (HCF), you are finding the largest number that divides a whole number of times into your original numbers. You might be asked about how many people there are when things are shared equally between them, or how many bags should be bought.

Example 3.5

An ice-cream company stores ice-cream in tubs that are cubes so they fit perfectly into large, rectangular chest freezers as shown.

What is the largest size that these cubes could be and how many would fit into the freezer at that size? *(4 marks)*

1.4 m

1.12 m

4.2 m

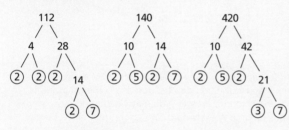

Freezer has dimensions 112 cm, 140 cm and 420 cm.
Prime factor decomposition:

As the side length of the cubes must be a factor of all the dimensions of the freezer, you need to find the HCF of 1.12, 1.4 and 4.2. HCF works for integers only so step one is to convert the dimensions into cm.

$112 = 2 \times 2 \times ② \times ② \times ⑦$
$140 = ② \times ② \times 5 \times ⑦$
$420 = ② \times ② \times 3 \times 5 \times ⑦$ ✓

The highest common factor is $2 \times 2 \times 7 = 28$ cm

It is good practice to place the factors in ascending order. That way it is easier to compare lists. Identify the prime factors that are 'in common'; that is the ones that appear for all three numbers.

The cubes could measure 28 cm in each dimension. ✓
In the freezer they would be arranged 4 deep by 5 high by 15 along. ✓
$4 \times 5 \times 15 = 300$ ✓
They could fit 300 tubs into the freezer at this size.

3.6 Indices

Using **powers** (or **indices**) to express a number can open up possibilities in the way you approach a question. You need to be able to interpret and evaluate indices that are fractions and negatives.

If you know there is a rule with indices but are struggling to remember it, try thinking about a simple case and what it means:

$$5^2 \times 5^3 = 5 \times 5 \times 5 \times 5 \times 5 = 5^5$$

You can see that when you multiply powers with the same roots, you can **simplify** by adding the powers.

Number

Example 3.6 📱

Simplify the following.

a) $2^{-4} \times 2^{-5} \times 2$ *(1 mark)*

$$2^{-4} \times 2^{-5} \times 2^1 = 2^{(-4 + (-5) + 1)} = 2^{-8} \checkmark$$

> Remember that a number without a power is that number to the power 1. When dealing with indices, writing in this 1 can sometimes help.

b) $3^{\frac{1}{2}} \times 3^{-\frac{3}{7}} \times 3^4$ *(2 marks)*

$$3^{\frac{1}{2}} \times 3^{-\frac{3}{7}} \times 3^4 = 3^{\left(\frac{1}{2} + \left(-\frac{3}{7}\right) + 4\right)} \checkmark$$

$$3^{\frac{1}{2}} \times 3^{-\frac{3}{7}} \times 3^4 = 3^{\frac{57}{14}} \checkmark$$

Working

$$\frac{1}{2} + \left(-\frac{3}{7}\right) + \frac{4}{1} = \frac{7}{14} - \frac{6}{14} + \frac{56}{14}$$

$$= \frac{7 - 6 + 56}{14} = \frac{57}{14}$$

> When adding fractions, find a common denominator.

> A fraction that is an **index** should be left improper (top heavy) and not turned into a mixed number.

> LCM of 2, 7 and 1 is 14. If you can find the LCM easily, it will save time later but if you don't spot it any common multiple will get you the right answer (you may need to do more steps in simplification and need to calculate larger numbers though).

c) $\dfrac{5^{-4} \times 5^6}{5^{\frac{2}{3}}}$ *(2 marks)*

$$\frac{5^{-4} \times 5^6}{5^{\frac{2}{3}}} = 5^{-4 + 6 - \frac{2}{3}} = 5^{\frac{4}{3}} \checkmark$$
$$\checkmark$$

> In this case you can either consider the fraction as a division or use a negative in the index. Both will get the same result.
> $$\frac{5^{-4} \times 5^6}{5^{\frac{2}{3}}} = 5^{-4} \times 5^6 \div 5^{\frac{2}{3}} = 5^{-4} \times 5^6 \times 5^{-\frac{2}{3}}$$

d) Evaluate $\left(\dfrac{16}{81}\right)^{\left(-\frac{3}{4}\right)}$ *(3 marks)*

$$\left(\frac{16}{81}\right)^{\left(-\frac{3}{4}\right)} = \left(\frac{81}{16}\right)^{\left(\frac{3}{4}\right)} \checkmark$$

$$\left(\frac{81}{16}\right)^{\left(\frac{3}{4}\right)} = \left(\frac{\sqrt[4]{81}}{\sqrt[4]{64}}\right)^{(3)} = \left(\frac{3}{2}\right)^{(3)} \checkmark$$

$$\left(\frac{3}{2}\right)^{(3)} = \left(\frac{3^3}{2^3}\right) = \frac{27}{8} \checkmark$$

> 'Evaluate' means 'find the answer to'. Here the index is both negative and a fraction. You can break it into three steps and deal with each element at a time. You can do them in any order but it makes sense to deal with the denominator before the numerator to avoid having to root a relatively large number.
> Step 1: The negative. Use the **reciprocal** of the number (flip it upside-down).
> Step 2: The denominator. This requires finding the root of the number; in this case you need to find the fourth root.
> Step 3: The numerator. Raise the number(s) to the power.

3.7 Accurate Answers – Surds and Pi

Surds are numbers expressed as a root of another number. Surds can accurately express **irrational numbers** (that is, numbers that cannot be written as a fraction, ones with decimals that never recur). If you are asked to give an accurate answer, it may involve surds or another form of irrational number, like π. If you find a decimal answer, even if you write all the digits on your calculator it will be a rounded answer and so not 100% accurate.

Example 3.7

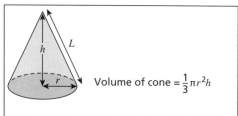

Volume of cone $= \frac{1}{3}\pi r^2 h$

A cone has a radius of 3 cm and a height of $\sqrt{48}$ cm.

a) Find the volume of the cone, giving your answer accurately and fully simplified.
(3 marks)

This example contains elements of geometry and algebra. The geometry, in fact, is just a context. The algebra required is a low level skill – **substituting** into a formula. The real skill tested in this question is your ability to express answers in surd format.

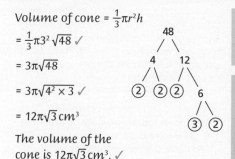

Volume of cone $= \frac{1}{3}\pi r^2 h$

$= \frac{1}{3}\pi 3^2 \sqrt{48}$ ✓

$= 3\pi\sqrt{48}$

$= 3\pi\sqrt{4^2 \times 3}$ ✓

$= 12\pi\sqrt{3}\ cm^3$

The volume of the cone is $12\pi\sqrt{3}\ cm^3$. ✓

Prime factor decomposition can help to find the largest square factor of a given number, which then helps you simplify the surd.

Conventionally the numbers (expressed in standard digits) are put at the beginning and it makes sense to put the surd at the end so there is no confusion about what is under the root sign and what is not, e.g. $\sqrt{3}\pi$ or $\sqrt{3\pi}$.

There is a square-based pyramid with the same height and volume as the cone.

b) Find the dimensions of the base of the pyramid. Give your answer accurately and fully simplified. *(3 marks)*

Often, solving part **b)** will rely on the answer from part **a)**. In this case you can work from the final answer or you can take one step backwards given that the heights are equal. The calculation can be simplified as shown below:

$$\frac{1}{3}b^2h = 12\pi\sqrt{3}$$

$$b^2\sqrt{48} = 3 \times 12\pi\sqrt{3} \checkmark$$

$$b^2 = \frac{36\pi\sqrt{3}}{\sqrt{48}} = \frac{36\pi\sqrt{3}}{4\sqrt{3}} \checkmark$$

$$b^2 = 9\pi$$

$$b = \sqrt{9\pi} = 3\sqrt{\pi}$$

The base of the pyramid measures $3\sqrt{\pi}$ cm by $3\sqrt{\pi}$ cm. $\checkmark$

$$\frac{1}{3}\pi r^2 h = \frac{1}{3}b^2h$$

$$\pi r^2 = b^2$$

$$b^2 = \pi 3^2 = 9\pi$$

$$b = \sqrt{9\pi} = 3\sqrt{\pi}$$

3.8 Surds – Rationalising the Denominator

Within mathematics there are certain conventions. Simplifying numbers is one of these, to make it easier to understand and compare. Rationalising the denominator is part of the simplification process. The convention is that you should not leave a number with an irrational denominator. It is also useful if you need to add or subtract the fractions.

Example 3.8

a) Simplify fully $\frac{\sqrt{24} + \sqrt{384}}{\sqrt{525}}$ *(3 marks)*

$$\frac{\sqrt{24} + \sqrt{384}}{\sqrt{525}} = \frac{\sqrt{2 \times 2 \times 2 \times 3} + \sqrt{2 \times 3 \times 8 \times 8}}{\sqrt{3 \times 5 \times 5 \times 7}}$$

$$= \frac{2\sqrt{2 \times 3} + 8\sqrt{2 \times 3}}{5\sqrt{3 \times 7}}$$

$$= \frac{10\sqrt{2}\sqrt{3}}{5\sqrt{3}\sqrt{7}}$$

$$= \frac{2\sqrt{2}}{\sqrt{7}} \checkmark$$

$$= \frac{2\sqrt{2}}{\sqrt{7}} \times \frac{\sqrt{7}}{\sqrt{7}} = \frac{2\sqrt{2}\sqrt{7}}{7} = \frac{2\sqrt{14}}{7} \checkmark$$

Where there are surds, start by simplifying each surd, finding any square factors and removing them from the surd. Prime factor decomposition can help if you are struggling to spot the factors.

At this point the denominator is irrational. You need to multiply the denominator by $\sqrt{7}$ to make the denominator **rational**. The numerator must also be multiplied by $\sqrt{7}$ so that the value of the number doesn't change (as $\frac{\sqrt{7}}{\sqrt{7}} = 1$).

b) Rationalise the denominator of $\frac{\sqrt{2}-\sqrt{3}}{\sqrt{3}-\sqrt{2}}$ *(3 marks)*

$$\frac{\sqrt{2}-\sqrt{3}}{\sqrt{3}-\sqrt{2}} = \frac{\sqrt{2}-\sqrt{3}}{\sqrt{3}-\sqrt{2}} \times \frac{\sqrt{3}+\sqrt{2}}{\sqrt{3}+\sqrt{2}} \checkmark$$

$$= \frac{\sqrt{6}+2-3-\sqrt{6}}{3+\sqrt{6}-\sqrt{6}-2} = \frac{-1}{1} \checkmark$$

$$= -1 \checkmark$$

Numerator

$(\sqrt{2}-\sqrt{3})(\sqrt{3}+\sqrt{2})$

$\sqrt{6}+2-3-\sqrt{6} = -1$

Denominator

$(\sqrt{3}-\sqrt{2})(\sqrt{3}+\sqrt{2})$

$3+\sqrt{6}-\sqrt{6}-2 = 1$

For more complicated denominators, where there are two terms, the 'difference of two squares' is used.

$(a - b)(a + b) = a^2 - b^2$

If either, or both, a and b are square roots (surds), $a^2 - b^2$ will be a rational number.

If you spot a shortcut, try to use it. Here, if you spot that the denominator can be factorised, so that one of the factors is the denominator, you can simplify.

$$\frac{\sqrt{2}-\sqrt{3}}{\sqrt{3}-\sqrt{2}} = \frac{\sqrt{2}-\sqrt{3}}{-(\sqrt{2}-\sqrt{3})} = -1$$

3.9 Accuracy and Bounds

Often the numbers dealt with are only accurate to a certain extent. This might be because rounding has happened to simplify the number being presented, or could be because a measurement is only accurate to half the smallest division of the scale. For example, if measuring a line with a ruler you can give an answer that is accurate to the nearest mm, possibly even half mm, but it would be very difficult, without a better scale, to be more accurate than that.

Example 3.9

In a speed test, a car accelerates from rest to a velocity of 60 miles per hour (accurate to the nearest 0.5 miles per hour). There are 1609.3 metres (accurate to 1 d.p.) in a mile. It takes the car 2.53 seconds (to 2 d.p.).

a) Find the upper and lower bounds of the speed of the car in metres per second (m/s) based on the information given. *(4 marks)*

Watch out for cases where the rounding isn't just to a certain digit (s.f. or d.p.). In this case the speed is accurate to the nearest 0.5 mph.

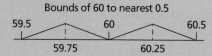

Bounds of 60 to nearest 0.5

So the **lower bound** is 59.75 and the **upper bound** is 60.25.

	Lower bound	Upper bound
Velocity	59.75	60.25 ✓
Metres in a mile	1609.25	1609.35 ✓

Find the upper and lower bounds of each given value first. Then consider the calculation. In this case, you are multiplying the values so taking the upper bound of each will give you the upper bound of the answer. If you are unsure, try different combinations to check, e.g. $60.25 \times 1609.25 \div 3600 = 26.9325...$ This falls between your values so isn't a bound.

To convert mph to m/s:
- Divide by 3600 (hours into seconds)
- Multiply by the number of metres in a mile

Upper bound = $60.25 \times 1609.35 \div 3600 = 26.934260416$ m/s ✓
Lower bound = $59.75 \times 1609.25 \div 3600 = 26.709079861$ m/s ✓

You can store these values in your calculator's memory so that if you need them in part b) you have them accurately.

b) By considering bounds, find the value of the car's acceleration to a suitable degree of accuracy. Give a reason for your answer. *(7 marks)*
Use the formula $a = \frac{v-u}{t}$

v is final velocity
u is initial velocity
a is acceleration
t is time taken

	Lower bound	Upper bound
Velocity (v)	26.709079861	26.934260416
Time (t)	2.525	2.535 ✓
Initial velocity	0	0

As you are dividing, the greatest possible value will come from an upper bound being divided by a lower bound. If the u value also had bounds, you would need to consider those. To get the biggest numerator you would do: upper bound of v – lower bound of u. A common mistake is to use all upper bounds in order to calculate the upper bound even when there is a subtraction or division.

Calculating the upper bound:
$a = \frac{v-u}{t} = \frac{26.9342... - 0}{2.525}$ ✓
$a = 10.667033828 ...$ ✓

Calculating the lower bound:
$a = \frac{v-u}{t} = \frac{26.7090... - 0}{2.535}$ ✓
$a = 10.5361261 ...$ ✓

The acceleration of the car was 11 m/s² accurate to 2 significant figures. ✓

The final degree of accuracy should be the number of significant figures for which both upper and lower bounds have the same answer. In this case an answer to 3 s.f. would give different answers (upper = 10.7, lower = 10.5) but the answer correct to 2 s.f. is 11 for both.

Reason: If you rounded to any more significant figures, the answer would be different for the upper and lower bounds (as supported by working above). ✓

Break down complicated calculations into components. For example P = $\left(\frac{a-b}{c}\right)d$, where a, b, c and d are all values with bounds. Work backwards through the order of operations.

- To get the largest value of P, you want the <u>upper bound of d</u> multiplied by the largest possible value of $\left(\frac{a-b}{c}\right)$.
- To get the largest possible value of $\left(\frac{a-b}{c}\right)$, you need the <u>lower bound of c</u> and the largest possible value of $a - b$.
- To get the largest possible value of $a - b$, use the <u>upper bound of a</u> and the <u>lower bound of b</u>.

3.10 Limits of Accuracy – Error Intervals

Truncation and rounding can be used to simplify an answer. You use them all the time. For example, if you are asked your age, you may truncate it to the nearest year. Sometimes the inherent inaccuracy of measurement means that a value is rounded. For example, if you draw a 14mm long line using a standard ruler, it is likely to be 14mm to the nearest half millimetre. More accurate instruments can reduce the error but even then the measurement won't be completely accurate. If a number is truncated or rounded, it can be important to know what the greatest and least value could be.

Example 3.10

Suz is cycling in a velodrome. She is training for a long distance ride, so she does 100 laps at a constant speed. It takes Suz 25 seconds, to the nearest second, to complete each lap.

a) What is the **error interval** for Suz's lap time? *(2 marks)*

25 seconds to nearest second

1 ÷ 2, split 0.5 seconds

$24.5 \leqslant t < 25.5$ ✓ ✓

Suz's coach says that she might not have time to complete 100 laps as she only has 40 minutes left. Her coach has truncated the time remaining to 2 significant figures.

b) What is the error interval for the time that Suz has left (T)? *(2 marks)*

$40 \leqslant T < 41$ ✓ ✓

As the number has been truncated, the stated value forms the lower bound (bottom of the error interval). The number could be 40.99999999 but truncate to 40 (truncated to 2 s.f.).

c) What is the error interval for the number of full laps that Suz will complete in her remaining time? *(4 marks)*

Most: Use her lowest lap time
and the longest time on track.

$41 \times 60 = 2460$ seconds

$2460 \div 24.5 = 100.4081633...$ ✓

Least: Use her highest lap time
and the lowest time on track.

$40 \times 60 = 2400$ seconds

$2400 \div 25.5 = 94.11764706...$ ✓

$94 \leqslant laps \leqslant 100$ ✓ ✓

As the question asks for the number of complete laps, you need to round down both your upper bound and your lower bound. Both these values are possible so both have the 'less than or equal to' inequality symbol.

3.11 Contextual Number Problems

Many number skills might be tested through a contextual problem. It can feel a bit like reading a story. Your skills lie in spotting the mathematical elements in the question and translating these into a series of calculations.

Example 3.11

Karen works for a factory that makes cake decorations. One of the products is chocolate star sprinkles. Every kilogram of chocolate makes one thousand stars. The factory sells the stars in packets of 120. Each kilogram of chocolate costs £2.50. The packaging for each packet costs 20p. The packets of stars are sold for £1.50 each. In a day the factory uses 225 kg of chocolate.

Assuming all packets produced are sold, how much profit will the factory make in one day? *(6 marks)*

Highlighting key information can help you to get the details from the question.

Number of stars made in one day = Amount of chocolate × 1000

$$= 225 × 1000$$

$$= \underline{225\,000 \text{ stars a day }} ✓$$

Cost of chocolate = Amount of chocolate × Cost of chocolate = 225 × 2.50

$$= 225 × 2 + 225 × \frac{1}{2} = 450 + 112.5$$

$$= \underline{£562.50 \text{ (cost of chocolate) }} ✓$$

Packets made each day = Number of stars ÷ Number of stars in a packet

$$= 225\,000 ÷ 120 = 22\,500 ÷ 12$$

$$= \underline{1875 \text{ packets made each day }} ✓$$

12	1
24	2
36	3
48	4
60	5
72	6
84	7
96	8
108	9
120	10

$$\begin{array}{r}0\;1\;8\;7\;5\\ 12\overline{)2^2 2^8 5^4 0^6 0}\end{array}$$

Sale price of packets of stars = Number of packets × Price

$$= 1875 × 1.50 = 1875 + 1875 × \frac{1}{2} = 1875 + 937.5$$

$$= \underline{£2812.50 \text{ income per day }} ✓$$

Cost of packaging = Number of packets × Cost

$$= 1875 × 0.2 = 1875 × 2 ÷ 10$$

$$= 3750 ÷ 10$$

$$= \underline{£375.00 \text{ (cost of packaging) }} ✓$$

$$\begin{array}{r}2\overset{7}{8}12.50\\ -\;\;562.50\\ \hline 2250.00\end{array}$$

Profit = Income – Outgoings (cost of chocolate and packaging)

$$= 2812.50 - 562.50 - 375.00$$

$$= \underline{£1875.00 } ✓$$

$$\begin{array}{r}\overset{1\;1\;\;\,1}{2}\overset{\;\,4\,1}{2}50.00\\ -\;\;375.00\\ \hline 1875.00\end{array}$$

Plan the steps to your answer:
There is no correct order. By explaining what each calculation does, you will be able to use the information you find more easily. The answer below has all the same steps and the same final answer but is much harder to understand. It also uses the = incorrectly.

$$225 × 1000 = 225\,000 ÷ 120 = 1875$$
$$225 × 2.5 = 562.5 + 20p × 1875$$
$$= 937.5$$
$$1875 × 1.5 = 2812.5 - 937.5$$
$$= 1875$$

 For more on the topics covered in this chapter, see pages 8–11, 24–27, 50–51 & 94–95 of the Collins Edexcel Maths Higher Revision Guide.

Number: Key Notes

- Integers are whole numbers including zero and negatives. Multiples and factors are part of 'whole number theory', so you will be working with integers.
- When comparing numbers in different forms, you need to be able to convert between fractions (F), decimals (D) and percentages (P).

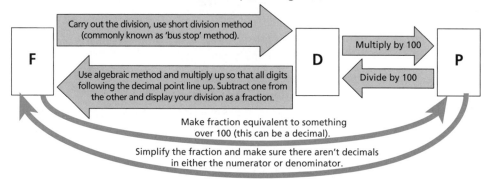

- There are some **sets** of numbers with special properties which can help you to answer questions. It is useful to be able to recognise the numbers as part of these sets:
  - **prime numbers:** 2, 3, 5, 7, 11, 13, 17, 19, 23, 29, …
  - **square numbers:** 1, 4, 9, 16, 25, 36, 49, 64, 81, 100, …
  - **cube numbers:** 1, 8, 27, 64, …, 1000, …
- Indices and **roots** are important and you need to be able to 'evaluate' and 'simplify' numbers given with fractional and negative indices.
- Standard form is used to express very large and very small numbers clearly. The form is having a single non-zero digit in front of the decimal point, then $\times 10^a$. For example, you can write 0.00000000103 as 1.03×10^{-9}.
- If asked for an accurate answer, you may be left with a surd or other irrational element in your number (e.g. π).
- To rationalise the denominator:
  - with a single surd in the denominator, multiply both the numerator and denominator by that surd.
  - with an **expression** in the denominator, e.g. $\frac{1}{a + \sqrt{b}}$, multiply both the numerator and denominator by $a - \sqrt{b}$.
- When working with bounds, find the bounds of all values, both upper and lower. In a calculation, carefully consider what combination of upper and lower bounds will give the greatest/least value. Break down complicated calculations into parts.
- With worded problems, the skill being tested is your understanding of what operations are needed to find the solution. Highlight or underline key information from the question and plan your answer, so you know what you are doing next at each stage. Show as much working as you can, making it clear and easy to follow.

4 Algebra

Algebra helps mathematicians to simplify the real world so that problems can be looked at objectively and more easily solved. Algebra makes up the largest proportion (27–33%) of the higher tier papers.

4.1 Manipulating Algebraic Expressions

Being able to confidently manipulate algebraic expressions is fundamental to success in algebra. It allows you to use different forms to draw out information.

Algebraic Fractions

Example 4.1.1

Which of the following are equivalent to $a^2 + 2ab$? Show working out to justify your answer. (4 marks)

$$\frac{4a^2 + 2ab}{4} \qquad \frac{a^2}{2} + \frac{2a^2 + 3ab}{3} \qquad \frac{3}{2}a\left(\frac{2}{3}a + \frac{4}{3}b\right) \qquad \frac{4a^2 + 8ab}{4}$$

$\dfrac{4a^2 + 2ab}{4}$

$\dfrac{\cancel{2}(2a^2 + ab)}{\cancel{4}^2}$

$\dfrac{2a^2 + ab}{2}$

Not equal ✓

$\dfrac{a^2}{2} + \dfrac{2a^2 + 3ab}{3}$

$\dfrac{3a^2}{6} + \dfrac{2(2a^2 + 3ab)}{6}$

$\dfrac{3a^2 + 4a^2 + 6ab}{6}$

$\dfrac{7a^2 + 6ab}{6}$

Not equal ✓

$\dfrac{3}{2}a\left(\dfrac{2}{3}a + \dfrac{4}{3}b\right)$

$\dfrac{3}{2} \times \dfrac{2}{3}a^2 + \dfrac{3}{2} \times \dfrac{4}{3}ab$

$\dfrac{\cancel{6}}{\cancel{6}}a^2 + \dfrac{\cancel{12}^2}{\cancel{6}}ab$

$a^2 + 2ab$

Equal ✓

$\dfrac{4a^2 + 8ab}{4}$

$\dfrac{\cancel{4}(a^2 + 2ab)}{\cancel{4}}$

$a^2 + 2ab$

Equal ✓

> When simplifying algebraic fractions, factorise before cancelling.

> To add algebraic fractions, make the denominators the same.

> As the question says you must justify your answer, your method needs to be clear. For each expression you should simplify to check whether it is equal. Don't forget to state for each one whether or not it is equal.

Expanding Binomials

Binomials are expressions with two **terms** ('bi' means 'two' and 'nomials' refers to the terms). You will not be expected to expand the product of more than three binomials. There are different methods for multiplying out brackets and it can be tempting to do them in your head, but showing a clear method is the best way to get full marks.

Example 4.1.2

Express $(x + 5)(3x - 2)(x - 4)$ in the form $ax^3 + bx^2 + cx + d$. *(4 marks)*

Whichever method you use, the first step is to multiply out two of the brackets and simplify. It doesn't matter which two you start with.

Method 1

	$3x$	-2	
	$3x^2$	$-2x$	x
	$+15x$	-10	$+5$

$3x^2 + 13x - 10$

	$3x^2$	$+13x$	-10	
	$3x^3$	$+13x^2$	$-10x$	x
	$-12x^2$	$-52x$	$+40$	-4

$3x^3 + x^2 - 62x + 40$

Method 2

$(x + 5)(3x - 2)(x - 4)$

$(x + 5)(3x - 2)$
$x(3x - 2) + 5(3x - 2)$ ✓
$3x^2 - 2x + 15x - 10$
$3x^2 + 13x - 10$ ✓
$(x - 4)(3x^2 + 13x - 10)$
$x(3x^2 + 13x - 10) - 4(3x^2 + 13x - 10)$ ✓
$3x^3 + 13x^2 - 10x - 12x^2 - 52x + 40$
$3x^3 + x^2 - 62x + 40$ ✓

Now use the same method to multiply the **quadratic** by the final bracket.

When you are checking your answer, it is the +/– that often causes mistakes.

Factorising

Factorising is the process of finding one or more factors of an algebraic expression. Questions will usually ask you to fully factorise an expression, which means you should be looking to take out the highest factors.

Example 4.1.3 (with fractions)

Fully factorise the expression $\frac{4a^2b^2}{15c} + \frac{4a^3b^3}{9c^2}$ *(2 marks)*

This looks more complicated because it has fractions but the method is exactly the same. You are looking for numbers and letters that are factors of both terms.

$4a^2b^2\left(\frac{1}{15c} + \frac{ab}{9c^2}\right)$ ✓

$\frac{4a^2b^2}{3c}\left(\frac{1}{5} + \frac{ab}{3c}\right)$ ✓ *Fully factorised*

There is no need to do this all in one go. It might be helpful to look at the numerator and the denominator separately. Here the factors of the numerator are taken out first, then the factors of the denominator.

Example 4.1.4 (with double brackets)

Fully factorise $3x^2 + 10x - 8$. *(3 marks)*

When the **coefficient** of x^2 is greater than 1, some students will use a slightly different method to factorise. The most common method is to split the middle term as described here. If you use a different method, that is fine. Whatever method you use, multiplying out to check your answers ensures you have factorised correctly.

$3x^2 + 10x - 8$

First make sure your quadratic is in the form $ax^2 + bx + c$.

$3 \times -8 = -24$

Multiply together the coefficient of x^2 and the value of c to get ac.

Product of –24	Sum of +10
-4×6	no
-3×8	no
-2×12	yes

Look for two numbers that multiply to give ac and add together to give b. Writing out the factors of ac can help if you cannot see it straight away.

$3x^2 + 12x - 2x - 8$

Split the middle x term using the two numbers you just found.

$3x(x + 4) - 2(x + 4)$ ✓

$(3x - 2)(x + 4)$ ✓✓

Factorise the first and last pairs of terms.

$3x^2 + 12x - 2x - 8$

$3x^2 + 10x - 8$

The two brackets are the same so factorising gives you the final answer.

There is no mark for multiplying out the brackets but it is the most important step as this is how you know you have got it right!

4.2 Mathematical Language and Structures

Mathematics has its own language and conventions to ensure that anyone will be able to understand what is happening. At times this can seem a bit confusing but usually it is so that the meaning can be conveyed with as little writing as possible.

Proof

Classic questions for proof are to do with odd and even numbers but you can be asked to use any other areas of maths in proof. A proof is a mathematical argument in which you are trying to convince the reader that something is true. The key to a good proof is being systematic and writing down every step very clearly.

Algebra

Example 4.2.1

Prove that the sum of two odd numbers will always be an even number. *(3 marks)*

Let n and m be integers.
Let $2n + 1$ be the first odd number.
Another odd number can be $2m + 1$. ✓

$2n + 1 + 2m + 1$
$= 2n + 2m + 2$
$= 2(n + m + 1)$ ✓

2 is a factor therefore it must be an even number. ✓

When a question talks about odd or even numbers, the starting point is to define them algebraically. By definition an even number must be a multiple of 2, meaning that $2n$ must be even. You can then say that $2n + 1$ must be an odd number.

Don't forget to explain why the algebra you have done proves the given statement; this is the conclusion to your argument.

Functions

A **function** is like a machine where there are inputs and outputs. The function tells you the relationship between the inputs and outputs.

Example 4.2.2

The functions f, g and h are such that $f(x) = \frac{x-9}{2}$, $g(x) = x^2 + 9$ and $h(x) = 2x$.

Find the values of x for which $f^{-1}(x) = gh(x)$. *(5 marks)*

$y = \frac{x-9}{2}$
$2y = x - 9$
$x = 2y + 9$
$f^{-1}(x) = 2x + 9$ ✓

Work out $f^{-1}(x)$ and $gh(x)$ separately before combining.

$f^{-1}(x)$ is the inverse of $f(x)$. Rearrange to make x the subject and then rewrite as $f^{-1}(x)$ but with the y value changed to an x.

$gh(x) = (2x)^2 + 9$
$gh(x) = 4x^2 + 9$ ✓
$f^{-1}(x) = gh(x)$
$2x + 9 = 4x^2 + 9$ ✓

To find $gh(x)$ you are substituting $h(x)$ in for the x values in $g(x)$. It is $2x$ all squared so the brackets are important here.

$0 = 4x^2 - 2x + 9 - 9$
$0 = 4x^2 - 2x$
$0 = 2x(2x - 1)$

It is a quadratic so can be put $= 0$ and factorised to solve.

$2x = 0 \qquad 2x - 1 = 0$
$x = 0$ ✓ $\qquad x = \frac{1}{2}$ ✓

There are two solutions, which makes sense as the question asks for the values of x.

Formulae

A formula is an equation that enables you to convert or find a value using other known values. There are some formulae you will need to know and these can be found in Chapter 10. Other formulae will be given as part of the question if needed.

Example 4.2.3

A ball is dropped from a height of 12 m. Its initial velocity is 0 m/s and it travels with a constant acceleration of 10 m/s².

Find the time it takes to reach the ground. Give your answer to 1 decimal place.

You may use this formula:

$$s = ut + \frac{1}{2}at^2$$

Where a is constant acceleration, u is initial velocity, v is final velocity, s is displacement from the position when $t = 0$ and t is time taken. *(3 marks)*

$s = 12$ $12 = (0)t + \frac{1}{2}(10)t^2$ ✓

$u = 0$ $12 = 5t^2$

$v = X$ $t^2 = \frac{12}{5}$ ✓

$a = 10$ $t = \sqrt{\frac{12}{5}}$

$t = ?$ $t = 1.549$

                  $t = 1.5$ seconds ✓

> You should have seen SUVAT equations before and be comfortable using them but they will be given if needed. It is a good idea to write out which bits of information you have and which you are looking for before you start.

4.3 Equations and Inequalities

To be able to use algebra to understand the real world, you need to be confident setting up equations from information. This will often draw upon other areas of mathematics, in particular geometry and measures. Throughout the rest of this chapter, many of the examples will be put into contexts. It is important to think about the implications of the context to check whether your answer seems sensible. The questions might not always be completely realistic but if, for example, you get a negative age you know that there is a mistake to check.

Algebra

Solving Quadratics

Example 4.3.1 (factorising)

The hypotenuse of a right-angled triangle is 13 cm. The length of the second longest side is 7 cm longer than the length of the shortest side.

Find the length of the shortest side. *(4 marks)*

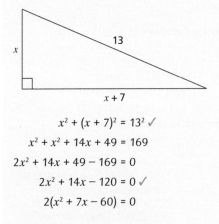

> The question is talking about a triangle, so a good starting point is to draw one and put on everything you know. You have been given three sides of a right-angled triangle, so you can use **Pythagoras' theorem**.

$$x^2 + (x + 7)^2 = 13^2 \checkmark$$
$$x^2 + x^2 + 14x + 49 = 169$$
$$2x^2 + 14x + 49 - 169 = 0$$
$$2x^2 + 14x - 120 = 0 \checkmark$$
$$2(x^2 + 7x - 60) = 0$$

$$x^2 + 7x - 60 = 0$$
$$(x + 12)(x - 5) = 0 \checkmark$$
$$x + 12 = 0 \qquad x - 5 = 0$$
$$x = -12 \qquad x = 5$$
$$x = 5\,cm \checkmark$$

> Taking out a factor of 2 makes the quadratic easier to factorise.

> Only one of the answers is sensible here so 5 cm is the answer.

Example 4.3.2 (using the formula)

A circular pond with a diameter of 8 m has a concrete area around it that is x m wide. Given that the concrete area is $10\pi\,m^2$, find the length of x giving your answer to 3 significant figures. *(5 marks)*

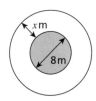

Diagram not to scale

x m

8 m

Large circle $\quad A = \pi(x + 4)^2$	Be clear which bits of information you are finding at each point by labelling the steps you are taking. Keep everything in terms of π for as long as possible as this is more exact and often values will cancel out.
$= \pi(x^2 + 8x + 16)$	
$= \pi x^2 + 8\pi x + 16\pi \checkmark$	
Small circle $\quad A = \pi 4^2$	
$= 16\pi \checkmark$	

Outside area $\quad \pi x^2 + 8\pi x + 16\pi - 16\pi$

$\pi x^2 + 8\pi x$

Before trying to solve the quadratic, you can divide through by π in order to simplify.

$\pi x^2 + 8\pi x = 10\pi \checkmark$

$\pi x^2 + 8\pi x - 10\pi = 0$

$x^2 + 8x - 10 = 0$

If you cannot see how to factorise the quadratic, you can either **complete the square** or use the quadratic formula.

$a = 1 \qquad b = 8 \qquad c = -10$

$\dfrac{-b \pm \sqrt{b^2 - 4ac}}{2a}$

$\dfrac{-8 \pm \sqrt{8^2 - 4(1)(-10)}}{2(1)} \checkmark$

$\dfrac{-8 \pm \sqrt{104}}{2}$

Write out the values of a, b and c as well as the formula before substituting in the values. This will mean you are less likely to make a mistake and, if you do make an error, you will be able to find it more quickly to make a correction.

$x = \dfrac{-8 + \sqrt{104}}{2} \qquad x = \dfrac{-8 - \sqrt{104}}{2}$

$= 1.09901... \qquad = -9.09901...$

$= 1.10 \qquad\quad x = -9.10$

$x = 1.10 \checkmark$

Check the degree of accuracy asked for in the question; the rounding here is a little tricky. Only one of these answers is possible for the value of x.

Linear Simultaneous Equations

Simultaneous equations will have two unknown variables which you need to find. There may be a question in which the equations are given and you are asked to find x and y, but often there will be a context with information that you need to use to form the equations.

Example 4.3.3

Boris and Amina have made cakes and biscuits to sell on a stall at the school fair to raise money. Amina sells her cakes for 30p each and her biscuits for 50p each. Boris charges 70p per cake and 50p per biscuit. Amina makes £8 and Boris makes £10. They both sell the same number of cakes and the same number of biscuits.

How many biscuits and cakes did they each sell? *(5 marks)*

Let c be the number of cakes sold.

Let b be the number of biscuits sold.

Amina (A)	Boris (B)
$30c + 50b = 800$	$70c + 50b = 1000$
$3c + 5b = 80$ ✓	$7c + 5b = 100$ ✓

The first thing to do with this question is to set up your equations. Begin by defining the variables. You could use any letters you like but it makes sense to use ones that fit with the context.

When setting up the equations, you need to check the units – the price of the cakes and biscuits is given in pence and the amount made is given in pounds. Both have been expressed in pence in the workings. This makes the numbers quite big. Both equations can be simplified by dividing through by 10.

$$B - A \quad 7c + 5b = 100$$
$$(-) \quad\quad 3c + 5b = 80 \checkmark$$
$$\frac{4c}{4} = \frac{20}{4}$$
$$c = 5 \checkmark$$

Substitute into A
$$3(5) + 5b = 80$$
$$15 + 5b = 80$$
$$5b = 65$$
$$b = 13 \checkmark$$

Check in B
$$7(5) + 5(13)$$
$$= 35 + 65$$
$$= 100$$

Because the coefficient of b is 5 in both equations, you can use elimination straight away. If you prefer to use substitution, that is fine; you will get the same answer.

Where you have a calculation like 13 × 5, it is a good idea to do it at the side of the page so that the examiner can see your method but it doesn't interrupt the flow of the algebra. There are no marks for the final check but it tells you that you have got it right and that is important for your peace of mind.

Quadratic Simultaneous Equations

A classic question on simultaneous equations is to find the points of **intersection** on a graph.

Example 4.3.4

The curve C with equation $y = x^2 - x - 12$ and the line L with equation $2x - y = 2$ intersect at the points A and B as shown.

Find the coordinates of A and B. *(5 marks)*

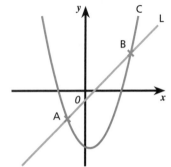

Rearrange L

$2x - y = 2$

$-y = 2 - 2x$

$y = 2x - 2$ ✓

Substitute into C

$2x - 2 = x^2 - x - 12$

$0 = x^2 - x - 2x - 12 + 2$

$0 = x^2 - 3x - 10$

$(x - 5)(x + 2)$

$x = 5 \qquad x = -2$ ✓

Find y values

at $x = 5$ at $x = -2$

$y = 2(5) - 2 \qquad y = 2(-2) - 2$

$= 8 \qquad\qquad = -6$ ✓

Check in C

$8 = 5^2 - 5 - 12 \qquad -6 = (-2)^2 - (-2) - 12$

$8 = 25 - 17 \qquad\qquad -6 = 4 - 10$

$B = (5, 8)$ ✓ $\qquad\qquad A = (-2, -6)$ ✓

In quadratic simultaneous equations, it is better to use substitution because elimination won't always work.

There are quite a few steps and things can become difficult to follow, so it is worth adding labels to show what you are doing at each stage.

The check tells you that you are correct, which is always a good feeling (and it will take less than a minute so don't be tempted to skip it).

It is nice at the end to write out the answer clearly in the context of the question.

Inequalities

Quadratic **inequalities** are different from linear inequalities as there may be more than one inequality needed. Students often have difficulties finding the final solution but sketching the curve will help you.

Example 4.3.5

Anna is making a run for her chickens. She has decided that the run should be rectangular with one side 1 m longer than the other. The chickens need a minimum area of 12 m² and she has a maximum of 22 m of fencing.

Find algebraically the possible lengths that the sides of the run could be. *(5 marks)*

Algebra

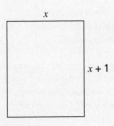

A good place to start is with a sketch.

There will be two inequalities, one for the perimeter and one for the area. Deal with the easiest one first.

Perimeter

$$x + (x + 1) + x + (x + 1) \leqslant 22$$
$$4x + 2 \leqslant 22$$
$$4x \leqslant 20$$
$$x \leqslant 5 \checkmark$$

Area

$$x(x + 1) \geqslant 12 \checkmark$$
$$x^2 + x \geqslant 12$$
$$x^2 + x - 12 \geqslant 0$$
$$(x + 4)(x - 3) = 0$$
$$x = -4, x = 3 \checkmark$$

Solving the quadratic tells you the roots of the quadratic, but you don't know the inequality or inequalities yet. A sketch is vital to get the answer right; don't be tempted to skip this step.

From the sketch you can see that there are two sections of the curve $y = x^2 + x - 12$ above $y = 0$. Only one of these is sensible as you cannot have a negative side length.

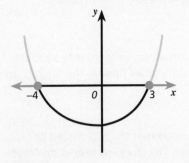

$$x \leqslant -4, x \geqslant 3 \checkmark$$

The two inequalities can now be combined. It can be helpful to draw a number line.

In metres, the side length must be $3 \leqslant x \leqslant 5 \checkmark$

4.4 Iteration

Iteration is the repetition of a process in order to find an approximate solution; the result of one iteration is used as the starting point for the next. Iteration can be used to generate sequences of numbers or to find approximate solutions to equations that could not otherwise be solved.

Example 4.4

a) Show that the equation $y = x^3 - 2x - 5$ has a root between 2 and 3. *(2 marks)*

At $x = 2$ $\quad y = (2)^3 - 2(2) - 5 = -1$ ✓
At $x = 3$ $\quad y = (3)^3 - 2(3) - 5 = 16$ ✓

The sign changes therefore the curve must cross the x-axis, meaning there is a root in this interval.

You need to state why your results mean that there is a root in the given interval.

b) Show that $0 = x^3 - 2x - 5$ can be rearranged into the form $x = \sqrt{2 + \frac{5}{x}}$ *(2 marks)*

$0 = x^3 - 2x - 5$
$2x + 5 = x^3$
$x^3 = 2x + 5$ ✓
$x^2 = \frac{2x}{x} + \frac{5}{x}$ ✓
$x^2 = 2 + \frac{5}{x}$
$x = \sqrt{2 + \frac{5}{x}}$

You have been given the answer so it is important to show your working out very clearly to get the marks.

If you get stuck on this part of the question, move on to the next part. You can always come back to it later and often the fresh perspective helps.

c) Starting with $x_0 = 2$ use the iteration formula $x_{n+1} = \sqrt{2 + \frac{5}{x_n}}$ to find an estimate for the solution of $y = x^3 - 2x - 5$ to 3 significant figures. *(3 marks)*

$x_0 = 2$
$x_1 = \sqrt{2 + \frac{5}{2}} = 2.1213...$ ✓
$x_2 = 2.0873...$
$x_3 = 2.0965...$
$x_4 = 2.0940...$ ✓
$x_5 = 2.0946...$
$x_6 = 2.0945...$
$x = 2.09$ to 3 s.f. ✓

Using a calculator: First type in the starting value and press 'enter' or '='. Now set up the equation with ANS where x_n is
$\sqrt{2 + \frac{5}{ANS}}$
Then continue to press '=' in order to generate the next solution.

From x_3 the value is the same to 3 s.f. The digit after could still mean it should round up and therefore change, so continue until the digit after is constant.

Algebra

4.5 Graphs

Linear Graphs

The general equation of a line graph is $y = mx + c$, where m is the **gradient** and c is the *y*-**intercept**.

Example 4.5.1

A line passes through the points A (0, 4) and B (4, 2) as shown in the diagram.
The line passing through the points B and D is perpendicular to the line passing through B and C.

Find the area of the triangle BCD. *(5 marks)*

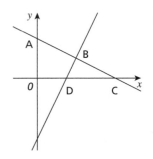

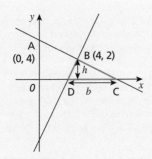

Adding everything you know on to the diagram can be very helpful with this sort of question. Drawing in the triangle can help you to see the pieces of information you already have and to identify what you need to work out.
To find the area of the triangle, you need to know the base and the height. The height is the *y*-coordinate of point B so is given in the question.
To find the base, you need to know the *x*-intercepts of both lines. These can be found from the equations of the lines.
When there are a lot of steps and more than one line, it can be easy to get confused so label any working out with which line you are working on.

Gradient of AB

$$m = \frac{2-4}{4-0} = \frac{-2}{4} = \frac{-1}{2}$$

Line AB $\quad y = \frac{-1}{2}x + 4$ ✓

$\qquad\qquad 0 = \frac{-1}{2}x + 4$

$\qquad\qquad \frac{1}{2}x = 4$

$\qquad\qquad\quad x = 8$ ✓

C is (8, 0).

For line AB you already know the *y*-intercept, so you can write down the equation of the line. Point C is the *x*-intercept of line AB and the *y*-coordinate is 0 so this can be substituted in to find the *x*-coordinate.

Gradient of BD

$m = 2$

Line BD $\quad 2 = 2(4) + c$ ✓

$2 = 8 + c$

$c = 2 - 8$

$c = -6$

So $y = 2x - 6$

At $y = 0$ $\quad 0 = 2x - 6$

$2x = 6$

$x = 3$ ✓

D is (3, 0).

The base of the triangle is $8 - 3 = 5$

$A = \frac{1}{2} \times b \times h$

$A = \frac{1}{2} \times 5 \times 2$

$A = 5$ units^2 ✓

> Line BD is perpendicular so the gradient is the negative reciprocal. You know the line goes through the point (4, 2), so you can use this to find the equation.

> With all of the information, you can now use the normal formula for the area of the triangle.

Graphs of Circles

Example 4.5.2

A circle (C) with centre O has the equation $x^2 + y^2 = 25$. Line 1 is a tangent to the circle at the point P (3, 4). Point A is the positive intersection of the circle C and the x-axis. Point B is the point of intersection of line 1 and the x-axis.

Find the length of AB. *(5 marks)*

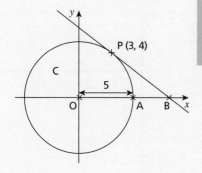

> A sketch will be really helpful here to see what is happening.
> The general equation of a circle is $x^2 + y^2 = r^2$, where r is the radius.

Gradient of OP = $\frac{4}{3}$ ✓

Gradient of the tangent = $\frac{-3}{4}$ ✓

$y = mx + c$

$4 = \frac{-3}{4}(3) + c$

$4 = \frac{-9}{4} + c$

$c = 4 + \frac{9}{4}$

$c = \frac{16}{4} + \frac{9}{4}$

$c = \frac{25}{4}$

$y = \frac{-3}{4}x + \frac{25}{4}$ ✓

When $y = 0$

$0 = \frac{-3}{4}x + \frac{25}{4}$

$\frac{3}{4}x = \frac{25}{4}$

$3x = 25$

$x = \frac{25}{3}$ ✓

$B - A = \frac{25}{3} - 5$

$\quad\quad = \frac{25}{3} - \frac{15}{3}$

$\quad\quad = \frac{10}{3}$ ✓

> To find point B you need to know the equation of the tangent which is perpendicular to the gradient of OP.

> You know that the y-coordinate of B is 0 because it lies on the x-axis, so you can substitute this in.

Sketching Curves

There are two types of questions concerned with sketching curves. You could be given the curve and asked to find the equation from the information given or you could be asked to sketch the curve from the equation.

Example 4.5.3 (using a curve)

The graph shows the quadratic f(x).

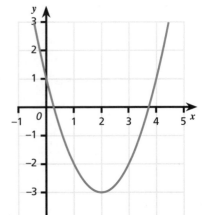

a) Write the equation of $y = f(x)$ in the form $y = (x + n)^2 + m$. *(2 marks)*

Turning point at $(2, -3)$

$y = (x - 2)^2 - 3$ ✓ ✓

The form you are being asked for is complete the square, which tells you the **turning point** of a quadratic. From the graph you can read off the coordinates of the turning point.

b) Find algebraically the coordinates of the points of intersection with the x-axis. Leave your answer in surd form. *(3 marks)*

When $y = 0$ $\quad (x - 2)^2 - 3 = 0$
$$(x - 2)^2 = 3 \quad ✓$$
$$x - 2 = \pm\sqrt{3}$$
$$x = 2 \pm \sqrt{3}$$

You could use the formula but you would have to multiply out the brackets first and it is already in complete the square form, so it is easier to carry on from here.

The points of intersection are at $(2 + \sqrt{3}, 0)$ and $(2 - \sqrt{3}, 0)$. ✓ ✓

Example 4.5.4 (sketching a curve)

a) On the same axes, sketch the curves of $y = x^3$ and $y = \frac{1}{x}$ *(2 marks)*

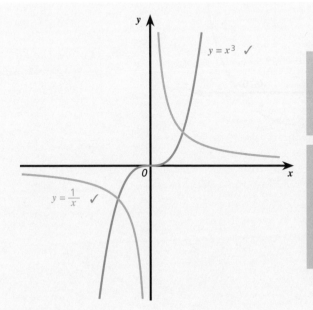

The reciprocal graph has **asymptotes** on both axes so make sure that your curves don't touch or go over the axes.

Because the graph of $y = x^3$ only has an intercept at the origin, there is no need to add any intercept labels. However, if it crossed at other points, they would need to be labelled.

b) Hence or otherwise, state the number of solutions to the equation $x^3 = \frac{1}{x}$. You must explain your answer. *(2 marks)*

> *The equation will have two solutions because the curves intersect twice.* ✓ ✓

'Hence or otherwise' means that part **a)** is probably going to be useful. You can do it another way but it makes sense to use the sketch in this case.

Transformations of Graphs

There are two types of transformations you need to know:
- Translations which can be in either the x or y direction
- Reflections in the coordinate axes.

Example 4.5.5

The graph of $y = f(x)$ is shown below.

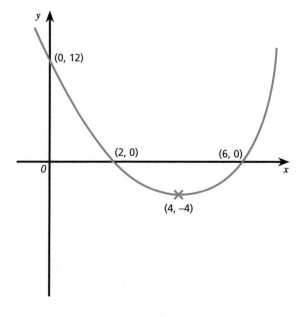

a) Sketch the graph of $y = -f(x)$. *(2 marks)*

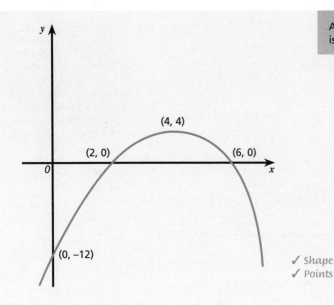

A transformation of $-f(x)$ is a flip in the x-axis.

(4, 4)

(2, 0)   (6, 0)

(0, −12)

✓ Shape
✓ Points

b) Sketch the graph of $y = f(x + 2)$. *(2 marks)*

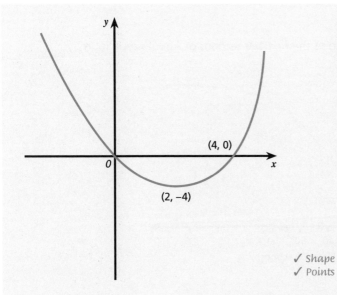

Translations in the x-axis go the opposite way to the direction you might expect.

(4, 0)

(2, −4)

✓ Shape
✓ Points

Algebra

c) Sketch the graph of $y = f(x) + k$ where $k > 4$. *(3 marks)*

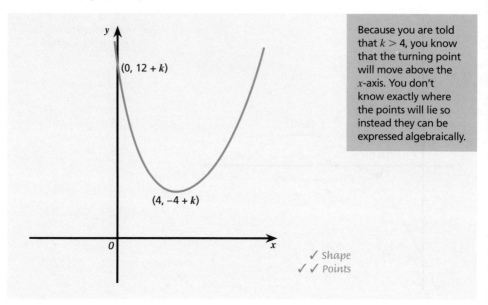

(0, 12 + k)

(4, −4 + k)

Because you are told that $k > 4$, you know that the turning point will move above the x-axis. You don't know exactly where the points will lie so instead they can be expressed algebraically.

✓ Shape
✓ ✓ Points

Gradients and the Area Under a Curve

Example 4.5.6

The speed–time graph shows the first 100 seconds of Abdul's cycle journey.

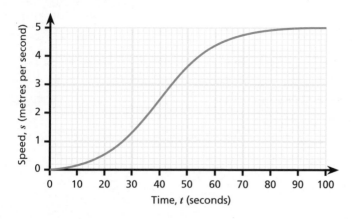

a) Find an estimate for the acceleration of the cyclist at the point $t = 25$. *(2 marks)*

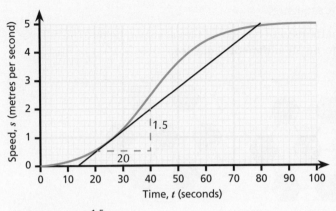

Draw the tangent to the curve at the point $t = 25$. The acceleration will be the gradient at this point.
To find the gradient, draw in a triangle and use $m = \dfrac{\text{Change in } y}{\text{Change in } x}$
Try to be as accurate as you can but there will be some differences so a range of answers will be given in the mark scheme.

Acceleration $= \dfrac{1.5}{20}$ ✓

$= 0.075\, m/s^2$ ✓

b) Work out an estimate for the distance Abdul travels in the first 60 seconds. Use three strips of equal width. *(3 marks)*

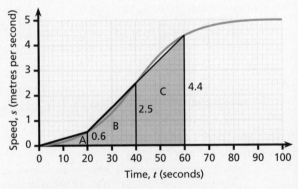

Distance is the area under the curve.

On the graph draw in the three sections that you will be using to estimate the area and label them.
Work out the area of each section separately and then add them together at the end.

A $\dfrac{1}{2} \times 20 \times 0.6 = 6$

B $\dfrac{1}{2}(0.6 + 2.5) \times 20 = 31$ ✓

C $\dfrac{1}{2}(2.5 + 4.4) \times 20 = 69$ ✓

A + B + C = 6 + 31 + 69 = 106 m ✓

Section A is a triangle because it starts at 0.
Sections B and C are trapezia with a height of 20.

4.6 Sequences

Example 4.6.1 (arithmetic sequence)

Anya is making mosaic patterns with hexagonal tiles as shown below. The number of tiles she uses each time forms an **arithmetic sequence**.

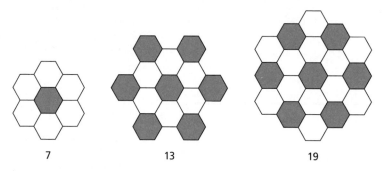

7   13   19

a) Can Anya make a pattern with 100 tiles? You must justify your answer. *(2 marks)*

7 13 19 $\bigvee \quad \bigvee$ 6 6	In this question you will need to justify your answer; just saying 'yes' or 'no' will not gain any marks.

nth term $6n + 1$

$6n + 1 = 100$

$-1 \quad -1$

$\dfrac{6n}{6} = \dfrac{99}{6}$

$n = \dfrac{33}{2}$ ✓

Find the nth term and put it equal to 100. If it is in the sequence, this will tell you which term it is.

n *is not an integer so 100 is not in the sequence.* ✓

State your conclusion clearly at the end.

b) Anya buys the tiles for 30p each. She spends £16.50 on tiles. Which term in the sequence will this be? *(2 marks)*

$\dfrac{1650}{30} = 55$ *tiles used* ✓

Start by working out how many tiles are used; remember to make sure the units are the same first.

$6n + 1 = 55$

$-1 \quad -1$

$\dfrac{6n}{6} = \dfrac{54}{6}$

$n = 9$

Put the nth term formula equal to the number of tiles and solve the equation.

The ninth term in the sequence will cost £16.50 ✓

c) Anya uses the formula £(3 + 1.5n) to decide how much to sell the mosaics for. She notices that this works well for small mosaics but she loses money on the larger pieces.

Which is the first term whereby Anya will lose money? *(3 marks)*

Cost of tiles in a mosaic is 0.3(6n + 1)	Using the nth term, write a formula for the cost of the tiles.
$\frac{0.3(6n + 1)}{0.3} > \frac{3 + 1.5n}{0.3}$ ✓ $6n + 1 > 10 + 5n$ $-5n \qquad -5n$	She will lose money when the cost of the tiles is greater than the price she is selling the mosaic for, so this can be written as an inequality.
$n + 1 > 10$ $-1 \quad -1$ $n > 9$ ✓	Solving the inequality shows that n must be greater than 9 so the tenth term will be the first to lose money.
The first she will lose money on is the tenth term. ✓	Make your answer clear at the end.

Example 4.6.2 (Fibonacci sequence)

Find the first term of a **Fibonacci sequence** in which the second term is 5 and the sixth term is 31. *(3 marks)*

Let the first term be x.

Term 1: x

Term 2: 5

Term 3: $x + 5$

Term 4: $5 + x + 5 = x + 10$ ✓

Term 5: $x + 5 + x + 10 = 2x + 15$ ✓

Term 6: $x + 10 + 2x + 15 = 3x + 25$

$3x + 25 = 31$

$3x = 6$

$x = 2$ ✓

Algebra

Example 4.6.3 (quadratic sequence)

A pattern is made from dots and lines as shown below.

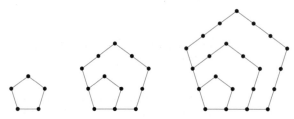

Find the number of dots used in the tenth pattern. *(4 marks)*

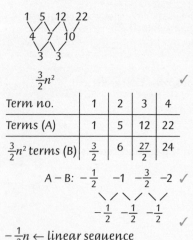

$$\frac{3}{2}n^2 \quad \checkmark$$

Term no.	1	2	3	4
Terms (A)	1	5	12	22
$\frac{3}{2}n^2$ terms (B)	$\frac{3}{2}$	6	$\frac{27}{2}$	24

$$A - B: \quad -\frac{1}{2} \quad -1 \quad -\frac{3}{2} \quad -2 \quad \checkmark$$

$$-\frac{1}{2} \quad -\frac{1}{2} \quad -\frac{1}{2} \quad \checkmark$$

$-\frac{1}{2}n \leftarrow$ linear sequence

$$\frac{3}{2}n^2 - \frac{1}{2}n$$

When $n = 1$

$$\frac{3}{2}(1)^2 - \frac{1}{2}(1) = \frac{3}{2} - \frac{1}{2} = 1$$

When $n = 2$

$$\frac{3}{2}(2)^2 - \frac{1}{2}(2) = \frac{12}{2} - \frac{2}{2} = 6 - 1 = 5$$

When $n = 10$

$$\frac{3}{2}(10)^2 - \frac{1}{2}(10) = \frac{300}{2} - \frac{10}{2} = 150 - 5 = 145$$

So there will be 145 dots in the tenth pattern. $\checkmark$

First look for the common difference between the terms. As it is changing each time, you then need to look for the second difference. The second difference divided by 2 tells you the coefficient of the n^2 term.

Find the value for each of the terms by substituting in the term number.

Subtracting the n^2 term from each of the original terms leaves you with a linear sequence. It looks tricky because of the fractions but don't be put off. It still works the same way.

Combine the linear and quadratic parts to get the nth term.

This question does not actually ask for the expression for the nth term, so if you find the answer by repetitive addition you will get the marks, <u>but only</u> if it is completely correct.

Check the answer is correct by substituting in some terms.

For more on the topics covered in this chapter, see pages 12–15, 38–41, 46–49, 66–73 & 96–101 of the Collins Edexcel Maths Higher Revision Guide.

Algebra: Key Notes

- To multiply out three binomials, start with a pair of brackets which can be multiplied out to give a quadratic and then multiply the quadratic by the final binomial.
- Completing the square rearranges a quadratic using the rule $x^2 + bx = \left(x + \frac{b}{2}\right)^2 - \left(\frac{b}{2}\right)^2$. If the coefficient of x^2 is > 1 then it needs to be factorised first.
- A proof is a mathematical argument, so it should show clearly the methods used and have a final statement to show why this has proved the point being made.
- An even number can be written as $2n$; an odd number can be written as $2n + 1$ or $2n - 1$.
- Functions show the relationships between inputs and outputs.
- fg(x) means that you are applying the function f to g(x) and should substitute in g(x) for each x.
- $f^{-1}(x)$ is the inverse of f(x). Rearrange to make x the subject and then rewrite as f(x) but with the y value changed to an x.
- Quadratics can be solved by factorising, completing the square or the quadratic formula $x = \frac{-b \pm \sqrt{b^2 - 4ac}}{2a}$.
- Linear simultaneous equations can be solved by elimination or substitution and will give a pair of solutions.
- Quadratic simultaneous equations can only be solved by substitution and will give two pairs of solutions.
- Inequalities are solved in the same way as equations but if you multiply or divide by a negative, you need to reverse the inequality sign.
- When solving quadratic inequalities, you will need a sketch to see which section or sections you need.
- Iteration is the repetition of a process in order to find an approximate solution; the result of one iteration is used as the starting point for the next.
- The general equation of a straight line is $y = mx + c$, where m is the gradient and c is the y-intercept.
- To find the gradient of a line, use $m = \frac{\text{Change in } y}{\text{Change in } x}$
- Parallel lines have the same gradient.
- The gradient of a perpendicular line will be the negative reciprocal $\frac{-1}{m}$
- When finding the equation of a line, substitute the value of the gradient and a pair of coordinates into $y = mx + c$ to find the value of c.
- The general equation of a circle with centre (0, 0) is $x^2 + y^2 = r^2$, where r is the radius.
- If asked to sketch a curve, make sure you include labels for all points of intersection with the axes.

Algebra

- You need to know the shape of these curves:

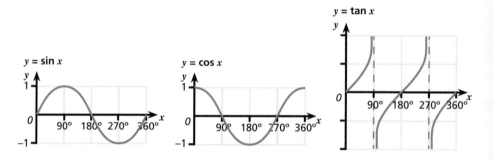

	Quadratic	Cubic	Reciprocal	Exponential
Positive	$y = x^2$	$y = x^3$	$y = \frac{1}{x}$	$y = 2^x$
Negative	$y = -x^2$	$y = -x^3$	$y = -\frac{1}{x}$	You don't need to know this one but would you be able to work it out from the others?

$y = \sin x$

$y = \cos x$

$y = \tan x$

- To find the turning point of a quadratic, rearrange into complete the square form where $y = (x + n)^2 + m$ and the coordinates of the turning point will be $(-n, m)$.
- A translation of f(x) + a moves f(x) up a; in other words, add a to the y-coordinates.
- A translation of f(x + a) moves f(x) to the left a; in other words, subtract a from the x-coordinates.
- A transformation of –f(x) is a flip in the x-axis and a transformation of f($-x$) is a flip in the y-axis.
- In a Fibonacci sequence, the previous two terms are added together to get the next term.
- For quadratic sequences, the second difference divided by 2 tells you the coefficient of the x^2 term. Subtract this from each of the terms so you are left with a linear sequence. Combine the two parts for the quadratic nth term and check that it works.

5 Ratio, Proportion and Rates of Change

Ratio, Proportion and Rates of Change (RPRoC) will be a relatively large proportion of the exam papers. 17–23% of marks will be based on this topic (approximately one-fifth of the papers).

Ratio is used to compare the parts within the whole and proportion compares each part to the whole. If you consider that your exam papers will be 20% RPRoC, that is proportion as you are considering how much of the whole. However, you could also say the papers will have 1 mark on RPRoC for every 4 marks on any other topic, giving a ratio of 1 : 4.

You need to be comfortable changing between the different forms of expressing ratios and proportions.

Rates of change look at the proportional increase, or decrease, to a population/value/quantity.

5.1 Scales – Maps, Models and Diagrams

Scales are often expressed as a ratio. The use of ratio means that there is no need for units of measurement (or that they can be used with any unit of measure). However, sometimes the ratio will be given in terms of 1 cm for every 5 m, or similar. You must ensure that your final answer makes sense and is in reasonable units. When establishing a **scale factor** or ratio between a model and real life, or between similar shapes, it is important to remember that the ratios between linear measurements need to be squared when applied to area, or cubed when applied to volume.

Example 5.1

Harriet is making some toy dinosaurs. Each is a scale version of the real dinosaurs, based on current expert opinion.

Dinosaur Toy				
	Tyrannosaurus Rex	Giraffatitan	Triceratops	Stegosaurus
Length	15 cm	20 cm	11 cm	9 cm
Height	5.8 cm	11 cm	3.4 cm	3.9 cm

Ratio, Proportion and Rates of Change

Each toy will have a tag with some key facts about the dinosaur attached to it. Harriet has started writing the labels but cannot remember which toy each label belongs to.

Being asked to justify your answer means you need to support your answer with good mathematical steps. A paragraph isn't necessary but some words can help to explain what you are doing.

a) Match each label to the right toy. Justify your answer. *(3 marks)*

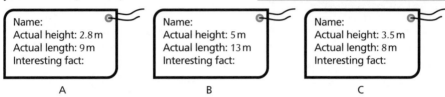

Name:
Actual height: 2.8 m
Actual length: 9 m
Interesting fact:

A

Name:
Actual height: 5 m
Actual length: 13 m
Interesting fact:

B

Name:
Actual height: 3.5 m
Actual length: 8 m
Interesting fact:

C

Dinosaur Toy	Tyrannosaurus	Giraffatitan	Triceratops	Stegosaurus
Length	15 cm	20 cm	11 cm	9 cm
Height	5.8 cm	11 cm	3.4 cm	3.9 cm
Length ÷ height	2.5862...	1.8181...	3.2352...	2.3076...

A: Length ÷ Height = 9 ÷ 2.8 = 3.2142...
This most closely matches the triceratops toy's length to height ratio (3.2352...).
B: Length ÷ Height = 13 ÷ 5 = 2.6
This most closely matches the t-rex toy's length to height ratio (2.5862...).
C: Length ÷ Height = 8 ÷ 3.5 = 2.2857...
This most closely matches the stegosaurus toy's length to height ratio (2.3076...). ✓
Label A goes on the triceratops, B goes on the tyrannosaurus rex and C goes on the stegosaurus. ✓

The final label has been more fully completed.

Name: Giraffatitan
Actual height:
Actual length: 22 m
Interesting fact: From the Jurassic period this dino held the largest dino record until the giant titanosaurians claimed the lead along with the sauroposeidon.

b) i) Express the scale, as a unit ratio, of the giraffatitan toy based on its label information. *(1 mark)*

Toy size : Actual size
20 cm : 22 m
20 : 2200
1 : 110 ✓

Be clear as to which way round you are establishing your ratio. Think whether there is a way that will make more sense or be neater if you are given a choice. Make sure that the units are the same.

ii) Find the missing height of the giraffatitan, rounded to the nearest half metre.
(2 marks)

The toy has a height of 11 cm.

11 × 110 = 1210 cm ✓

Height of actual giraffatitan = 12.0 m
(to the nearest half metre) ✓

Always look out for final answer form. In this case, the question has asked you to round to the nearest half metre. 12.1 m is between 12.0 m and 12.5 m but closer to 12, so you round down.

c) The volume of a life-sized tyrannosaurus rex's head is about 1.6 m³.
What volume should the head of the scale toy be? *(5 marks)*

There isn't a lot of information here, so you may need to reference earlier parts of the question.

Linear (one-dimensional) scale of model to real life

15 cm : 13 m

15 : 1300

$1 : 86\frac{2}{3}$ ✓

To convert this into a three-dimensional scale, cube it. ✓

1 : 650 962.962962...

We know that this needs to relate to

$y : 1.6\,m^3$

$1.6\,m^3 = 1.6 \times (100^3) = 1\,600\,000\,cm^3$ ✓

When you see volume or area in a scale question, make sure you don't apply the linear scale but make the necessary adjustment.

$\times\frac{5400}{2197}$ ⟋ $y : 1\,600\,000$
⟍ $1 : 650\,962.962962$ ⟍ $\times\frac{5400}{2197}$ ✓

$y = 1 \times \frac{5400}{2197} = 2.45789... = 2.5\,cm^3$ (1 d.p.)

Multiplier found by dividing 1 600 000 by 650 962.962.... If you got a different answer, check if you have rounded a number early, or if it is just in a different form. Use your calculator's memory to help store complicated numbers part way through calculations.

The tyrannosaurus rex toy should have a head with a volume of 2.5 cm³. ✓

Consider the units you choose for your final answer. In this case, by converting into a smaller unit, the final answer is in a form that makes sense. If you got an answer of $\frac{1}{400000}$ = 0.0000025 m³, it is harder to judge if that is a reasonable answer, even though it is exactly the same answer.

Ratio, Proportion and Rates of Change

5.2 Ratios

Ratios are used as a way of expressing quantities. They might be used for converting currencies or mixing colours of paint. There are lots of different ways you can be asked about ratio, some more obvious than others. Be clear what the question gives to you and what it is you are trying to find.

Example 5.2 🖩

Jenny and Ted are making drinks of fruit juice. Jenny likes to have her juice diluted in a ratio of juice to water of 3 : 1. Ted likes his juice to water ratio to be 1 : 1. There is 240 ml of juice left.

a) Jenny and Ted share the juice equally.
What is the difference in volume of their drinks once diluted? *(3 marks)*

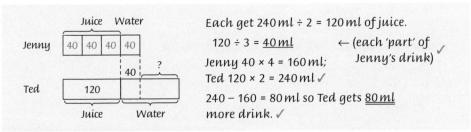

Each get 240 ml ÷ 2 = 120 ml of juice.

120 ÷ 3 = <u>40 ml</u> ← (each 'part' of Jenny's drink) ✓

Jenny 40 × 4 = 160 ml;
Ted 120 × 2 = 240 ml ✓

240 − 160 = 80 ml so Ted gets <u>80 ml</u> more drink. ✓

b) How much juice should they each start with so that the final volume of their drinks is equal? *(2 marks)*

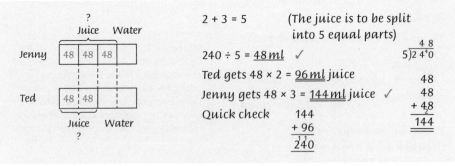

2 + 3 = 5   (The juice is to be split into 5 equal parts)

240 ÷ 5 = <u>48 ml</u> ✓

Ted gets 48 × 2 = <u>96 ml</u> juice
Jenny gets 48 × 3 = <u>144 ml</u> juice ✓

Quick check 144
+ 96
240

$$5)\overline{2\,4^40}$$ 48

48
48
+ 48
144

c) What fraction of Ted's drink is water? *(1 mark)*

$\frac{1}{2}$ ✓

1 : 1 means there are two parts altogether, of which one is water.

5.3 Ratios – Comparing and Calculating

Using unit ratios, or the form $1 : x$, can be a good way to compare different ratios.

Example 5.3

An artist is mixing colours for her painting. She used a mix of midnight-blue and white in a ratio of 5 : 2 to make a lighter blue. She now wants to make a slightly paler blue, to add highlights.

a) She chooses from the following ratios for the slightly paler blue:
8 : 3, 12 : 5 or 15 : 6

Which should she use? *(3 marks)*

To compare the ratios, you can turn them all into unit ratios, i.e. $1 : x$. To do this you divide both sides by the value of the left-hand side.

5 : 2 *(divide both sides by 5)* $\Rightarrow$ 1 : 0.4 $2 \div 5$

$\rightarrow$ $5\overline{)2.0}$ → 0.4

8 : 3 *(divide both sides by 8)* $\Rightarrow$ 1 : 0.375 $3 \div 8$

$\rightarrow$ $8\overline{)3.^{3}0^{6}0^{4}0}$ → 0.375

12 : 5 *(divide both sides by 12)* $\Rightarrow$ 1 : 0.41̇6 ✓ $5 \div 12$

$\rightarrow$ $12\overline{)5.0^{2}0^{8}0^{8}0...}$ → 0.4166...

15 : 6 *(divide both sides by 15)* $\Rightarrow$ 1 : 0.4 ✓ $6 \div 15$

$\rightarrow$ $15\overline{)6.0}$ → 0.4

12 : 5 is the ratio she should use. ✓

Once expressed in a way that is easier to compare, you need to be careful about your interpretation. This question is asking for a lighter colour, which means the amount of white for each part blue has to be higher.

b) The artist has lots of tubes of midnight-blue paint but only three tubes of white. She decides to use all of the white tubes of paint.
How much blue should she mix in to get the same paler colour as part a)? *(2 marks)*

Blue : White
$\times\frac{3}{5}$ ⌢ 12 : 5 ⌢ $\times\frac{3}{5}$
↘ x : 3 ↗

$12 \times \frac{3}{5} = 12 \times 3 \div 5$ ✓

$= 36 \div 5$

$= 36 \times 2 \div 10$

$= 72 \div 10$

$= 7.2$

She should mix in 7.2 tubes of blue paint to get the same colour. ✓

To find the multiplier, you can use the fact that division is the inverse operation. So $3 \div 5 = \frac{3}{5}$ gives the multiplier. This is then used to multiply both sides to find the equivalent ratio with three parts white.

There are many ways of carrying out calculations without your calculator; it is up to you which method you use.

Always check your answer makes sense and has units; in this case, the units are 'tubes of paint'. 7.2 is a reasonable number but check if using a decimal makes sense in the context; in this case, you can get 0.2 or one-fifth of a tube of paint.

c) Each tube contains 30 ml of paint. The artist decides to use five parts midnight-blue to every four parts of white.
To get the required texture, she adds sand to the paint in a ratio of 1 : 3.
Her mixture will be applied so that every 15 ml of mixture covers 100 cm^2.
She needs to cover 0.8 m^2 on her canvas.

Having used up all her supplies, how many tubes of each colour paint should she put on her shopping list, and how much sand? *(5 marks)*

This is another way in which the question could be asked. It is unlikely that all three parts of this question would appear in one exam question. Can you think of any other ways a similar question could be asked?

The ratio represents things in the order they are mentioned, unless explicitly stated otherwise. In this case, sand is first so is represented by the first number in the ratio.

Area needing to be covered = 0.8 m^2 = 0.8 × 100^2 = 8000 cm^2
Paint mix required = 15 × 8000 ÷ 100 = 15 × 80 = 1200 ml ✓
1200 ml split into the ratio 1 : 3
1200 ÷ 4 = 300, so each part is 300 ml. ✓
1 part sand, so she needs <u>300 ml of sand</u>.
3 parts paint, so she needs <u>900 ml of paint altogether</u>.
Paint is in the ratio of 4 : 5, so there are 9 parts altogether. ✓
900 ÷ 9 = 100
She will need 400 ml of white paint and 500 ml of blue paint.
400 ÷ 30 = 40 ÷ 3 = 13.33... ✓
500 ÷ 30 = 50 ÷ 3 = 16.66...
She should buy 14 tubes of white paint and 17 tubes of blue paint.
She needs 300 ml of sand. ✓

5.4 Compound Measure

Compound means a mixture of more than one measure. In this case, it is measures that are expressed as a combination of the standard base units. Pressure is Nm^{-2} or N/m^2. This is the force (measured in Newtons, N) divided by the area (measured in m^2). Dimensional analysis can be a really helpful way of supporting and checking your working. If your answer is in Nm^{-2}, then to get there you must divide the force by the area.

Example 5.4

Speed is a measurement of distance divided by time. Joanna is travelling at a constant speed on the motorway and the sat-nav tells her that it is 6.5 miles to her exit. She says that this will take her 10 minutes as she is travelling at 65 mph. Is she correct? *(2 marks)*

$Speed = 65\,mph$

$Distance\ to\ go\ is\ 6.5\ miles$

$s = \dfrac{d}{t}$

$t = \dfrac{d}{s}$

$t = \dfrac{6.5}{65} = 0.1\ hours\ \checkmark$

$t = 6\ minutes$

> When presented with a possible solution to a problem, you can be misled trying to follow the method. Try starting with the elements of the problem and finding your own answer to compare.

Joanna is incorrect. It will take her 6 minutes to reach her exit travelling at 65 mph. ✓

5.5 Direct Proportion – Recipes

Recipes are often used as a way of looking at proportions. The recipe will work as long as the proportions of ingredients stay the same. If you use 100 g flour for every egg, then for two eggs you would use 200 g flour. If you only had 50 g flour, you would use half an egg. There is also lots of possibility for unit conversions too, so watch out!

Example 5.5

Gerald is baking some buns for a cake sale at school. He wants to make as many as he can with the ingredients he already has.
He knows that 1 oz ≈ 28 g, and 1 pint = 32 tablespoons.

Gerald has:
- Plenty of food colouring and vanilla extract
- 1.5 kg of butter
- One and a half bags of self-raising flour (500 g when full)
- $\frac{3}{4}$ of a bag of caster sugar (1 kg when full)
- 2 kg of icing sugar
- $\frac{1}{2}$ pint of milk
- A dozen free-range eggs

To make 12 buns:
4 oz butter or margarine
4 oz self-raising flour
3 oz caster sugar
1 tsp vanilla extract
$1\frac{1}{2}$ tbsp milk
2 free-range eggs, lightly beaten

For the buttercream icing
5 oz butter, softened
10 oz icing sugar
1.5 tbsp milk
A few drops of food colouring

Ratio, Proportion and Rates of Change

a) What is the greatest number of buns Gerald can make whilst using a whole number of eggs? *(5 marks)*

> There are lots of ways to be clear in your working. Tables can help set out working where there is a process repeated a number of times.

> In this case, it does not matter if you convert the recipe into grams or the ingredient list into ounces. The numbers will be different but the theory remains the same.

For 12 buns Gerald needs			For one bun	He has	How many buns?
Butter	$(4 + 5) \times 28$	252 g	21 g	1500 g	71.428...
Self-raising flour	4×28	112 g	$9\frac{1}{3}$ g	750 g	80.357...
Caster sugar	3×28	84 g	7 g	750 g	107.142...
Icing sugar	10×28	280 g	$23\frac{1}{3}$ g	2000 g	85.714...
Milk	$1.5 + 1.5$	3 tbsp	$\frac{1}{4}$ tbsp	$0.5 \times 32 =$ 16 tbsp ✓	64
Eggs		2 eggs	$\frac{1}{6}$ egg	12 eggs	72 ✓✓

Buns Gerald can make (*b*)

$b \leqslant 64$ (as he is limited by milk) ✓

Each whole egg makes 6 buns so Gerald can make 60 buns using a whole number of eggs. ✓

> If you spot a shortcut, such as looking at multiples of half the recipe (which works in this case as that is for each whole egg), then you can use it. The method of finding 'one bun' then multiplying up will work in all cases. But you then need to consider the whole eggs at the end.

b) When baking, 15% of the buns found in part **a)** get burned and are thrown away. The rest Gerald ices in purple and red in a ratio of 8 : 9. He sells the purple buns in bundles of three for 50p and the red buns individually for 20p each.

If Gerald manages to sell all his buns, how much money will he make? *(4 marks)*

60 buns go into the oven

15% of 60 = 6 + 3 = 9

60 − 9 = 51 (buns to be sold) ✓

8 + 9 = 17

51 ÷ 17 = 3

Purple : Red

$8 \times 3 : 9 \times 3$

24 : 27   (Check 24 + 27 = 51) ✓

Income from purple buns is 24 ÷ 3 = 8

8 × 0.50 = £4 ✓

Income from red buns is 27 × 0.20 = £5.40

Total income = 4 + 5.40 = £9.40 ✓

5.6 Direct Proportion

In **direct proportion**, as one element increases the other does by the same proportion. So if one value doubles, so does the other. If y is proportional to x^2, then if y doubles the value x^2 will double too.

Example 5.6

A flower design is produced out of congruent regular hexagons. The area of each hexagon is proportional to the square of the length of its edge. A hexagon has an area of $6\,cm^2$ when its edge measures $\frac{2}{\sqrt[4]{3}}$ cm.

a) What is the area of a hexagon with side length 4 cm? Give your answer accurately. *(3 marks)*

$A \propto x^2$, where x is the length of an edge and A is the area.

$A = kx^2$

$6 = k \times \left(\frac{2}{\sqrt[4]{3}}\right)^2$ ✓

$k = \frac{6\sqrt{3}}{4} = \frac{3\sqrt{3}}{2}$ ✓

For the new situation, each hexagon has the area:

$A = \frac{3\sqrt{3}}{2} \times 4^2 = 24\sqrt{3}\,cm^2$ ✓

b) What is the perimeter of the design if the total area is $42\sqrt{3}\,cm^2$? *(4 marks)*

Total area $42\sqrt{3}$

$42\sqrt{3} \div 7 = 6\sqrt{3}$, area of each hexagon ✓

$A = kx^2$

$6\sqrt{3} = \frac{3\sqrt{3}}{2} \times x^2$ ✓

$x^2 = 6\sqrt{3} \div \frac{3\sqrt{3}}{2} = \frac{12\sqrt{3}}{3\sqrt{3}} = 4$ ✓

$x = 2$

$P = 18x = 18 \times 2 = 36\,cm$

Perimeter of the shape is 36 cm. ✓

Ratio, Proportion and Rates of Change

5.7 Inverse Proportion

Inverse proportion is when one value is proportional to the reciprocal of the other. If a is inversely proportional to b, then $a \propto \frac{1}{b}$.

Example 5.7

For a set length of wire, the resistance is inversely proportional to the **cross-sectional** area. The wire is cylindrical. When the thickness of the wire is 0.4 mm, the resistance is 60 Ω.

a)   Calculate the resistance if the wire had a thickness of 0.5 mm. *(3 marks)*

$R \propto \frac{1}{A}$ where $A = \pi r^2$

$R = \frac{k}{r^2}$ ✓

Using the given situation to find k:

$k = R \times r^2$

$\quad = 60 \times 0.2^2 = 2.4$ ✓

Using k to find the new situation with R:

$R = \frac{2.4}{0.25^2} = 38.4\,\Omega$ ✓

$R = \frac{k}{\pi r^2}$ is acceptable here but since both k and π are **constant** values you can combine them into a single **constant of proportionality**. Similarly the use of different units of measure will be absorbed by the constant, so as long as your units are consistent it is fine.

k is a constant so has no units.

b)   What must the thickness of the wire be to have a resistance of 15 Ω? *(3 marks)*

$R = \frac{k}{r^2}$

$r^2 = \frac{k}{R}$

$r = \sqrt{\frac{k}{R}}$ ✓

$r = \sqrt{\frac{2.4}{15}} = \sqrt{\frac{4}{25}} = 0.4$ ✓

The wire must be 0.8 mm thick to produce a resistance of 15 Ω. ✓

You can either rearrange algebraically (as has been done here) to make r the subject of the formula, or you can substitute in the values you know, then solve from there.

15 Ω is a lower resistance than the others so you would expect the wire to have a larger cross-sectional area. The value of r will be bigger and then you need to double it to get the thickness of the wire. In fact, the resistance is a quarter of the original situation, so the area will be four times as big, meaning the thickness will be double.

5.8 Growth and Decay

There are many key words associated with growth and decay. You need to recognise them and understand them. **Depreciation** means a decrease in the value and decay tends to mean a decrease in the quantity (and is generally given as a percentage). **Interest** means an increase in value and growth means an increase in the quantity (e.g. population size).

Example 5.8

Jaden bought a brand new car. In the first year, it depreciated by 38%. He predicts that after the initial drop it will depreciate steadily at a rate of 12.5% each year.

a) Jaden says that, as 38 + 12.5 = 50.5%, after the second year the value of the car will be just under half its original value.
Is Jaden correct? You must give a reason for your answer. *(2 marks)*

Jaden is incorrect as the 12.5% is only applied to the value of the car at the start of the second year. ✓

The car retains 62% of its value in year 1. It retains 87.5% of the value in the second year.

$0.62 \times 0.875 = 0.5425 = 54.25\%$ of the original value at the end of the second year. ✓

His car will be worth over half its original value.

A common mistake is to add the percentage values. Remember that the depreciation of 12.5% is applied once the 38% has already been taken off. It is not 12.5% of the original amount but 12.5% of the 62% that was there at the start of the second year. (12.5% of 62% = 7.75%, so it is actually a further 7.75% of the original amount).

b) Jaden bought the car for £10 155. He decides to sell it when the value is less than £3000. He says that it will be after seven years but his friend says it will be after six years.
Who is right? *(3 marks)*

After 6 years the value of the car will be
$10\,155 \times 0.62 \times 0.875^5 = 3229.325... = £3229.33$ ✓
After 7 years the value of the car will be
$10\,155 \times 0.62 \times 0.875^6 = 2825.6602... = £2825.66$ ✓
Jaden is correct, as after 6 years the value will still be above £3000. ✓

As the first year had a different depreciation rate, remember the power here is one less than the number of years.

Calculator shortcut: many calculators let you scroll back into the first calculation and change the numbers, or calculation, which saves typing it all in again. Get to know your calculator!

Ratio, Proportion and Rates of Change

5.9 Growth and Decay – Simple and Compound

Interest rates apply to many things. House sales and house values are key examples.

Example 5.9

Brook and Charlie bought a house costing £237 000. The value of the house increases by the same percentage each year and after three years the value of the house is £278 295.21.

a) What is the percentage increase in the value of the house each year? *(3 marks)*

This question requires you to work out a compound (repeated) percentage change. If you get stuck, think what you would do if you were given the interest rate and consider setting up an equation based on that.

$237\,000 \times x^3 = 278\,295.21$

$x^3 = 278\,295.21 \div 237\,000 = 1.17424139...$ ✓

$x = \sqrt[3]{1.17424139...} = 1.055000055...$ ✓

This is a proportional increase of 0.055, which is equivalent to 5.5% each year. ✓

Check $237\,000 \times 1.055^3 = 278\,295.21$ (2 d.p.)

This will find the decimal multiplier so remember to convert from that into a **percentage increase** or **decrease**.

Brook and Charlie paid an initial deposit of 23% and borrowed the rest from a bank on a fixed-rate mortgage. The bank charges a fixed-rate interest of 4% per annum on the borrowed money.

Brook and Charlie pay monthly mortgage instalments which are equivalent to paying £10 000 at the start of each year.

b) What proportion of the house do Brook and Charlie own after two years, in relation to its value at that time? Give your answer to a suitable degree of accuracy. *(6 marks)*

Initially borrowed from bank = 77% of initial price
= 0.77 × 237 000 = 182 490, B&C own the rest. ✓
After 1 year, B&C have paid £10 000.
Mortgage before interest added = 182 490 − 10 000 = 172 490 ✓
Mortgage after interest added = 172 490 × 1.04 = 179 389.60 ✓
After 2nd year, B&C paid a further £10 000.
Mortgage before interest added = 179 389.60 − 10 000 = 169 389.60
Mortgage after interest added = 169 389.60 × 1.04 = 176 165.184 ✓
The value of the house is 237 000 × 1.055² = 263 786.925
263 786.925 − 176 165.184 = 87 621.741 ✓
87 621.741 ÷ 263 786.925 = 0.33216862...
= 33.2% (3 s.f.) ✓

5.10 Algebraic and Graphical Representations

You can represent proportionality (both direct and inverse) using graphs. You could be asked to convert currency using a graph. You need to be able to recognise what the graph for each relationship would look like and be able to interpret values from the graphs.

Example 5.10

The graph shows the relationship between the base of a triangle (x) and its height (y) given a constant area (A).

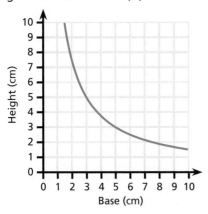

a) What is the value of A? *(1 mark)*

At $x = 3$, $y = 5$
so $A = \frac{1}{2}xy$

$= \frac{1}{2} \times x \times y$

$= \frac{1}{2} \times 3 \times 5 = 7.5\,cm^2$ ✓

b) What is the relationship between x and y in this situation? *(1 mark)*

x and y are inversely proportional. ✓

5.11 Compound Measure and Ratios

Often questions will ask you to decide which methods to use. Planning is important!

Example 5.11

Jirair has a cuboid made of solid copper.
This is a conversion graph for inches into cm.
Copper has a density of 8.96 g/cm³.
The ratio of the length to width to height of the cuboid is 5 : 1 : 3.
The height of the cuboid is 2.35 inches.

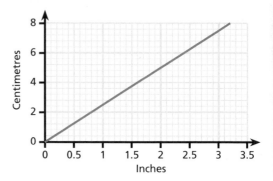

a) What is the mass of Jirair's copper cuboid? Give your answer in kilograms to 3 significant figures. *(5 marks)*

From the graph, 2.35 inches ≈ 6 cm ✓

$5 : 1 : 3 = 5x : x : 6$

$\qquad = 10 : 2 : 6$ ✓

Volume of the cuboid = 10 × 2 × 6 = 120 cm³ ✓

Mass of cuboid = 8.96 × 120 ✓

$\qquad\qquad = 1075.2\,g$

$\qquad\qquad = 1.0752\,kg = 1.08\,kg$ (3 s.f.) ✓

Don't be put off by things that can look quite complicated. Often if you start with something that you know, it can make everything become a lot clearer.

Is this the final answer? Is it in the required form?

As ever, there are different ways to approach this question. If you find the volume in inches, first remember the conversion ratio is the cubed linear measure. As part of the question is a reading from the graph, you may have a slightly different inches to centimetres conversion, especially if you read a different point. Be clear and, as long as your reading is reasonable, you should get the marks.

b) What is the approximate density of copper expressed in kg/inch³? Is this a sensible unit to use in this situation? *(5 marks)*

8.96 g/cm³ = 0.00896 kg/cm³ ✓

From the graph 2.35 inches ≈ 6 cm

Gives ratio of

2.35 : 6 = 1 : 2.553191... for lengths ✓

1 : 16.64371... for volumes ✓

0.00896 kg/cm³ = 0.14912764... kg/inch³

= 0.149 kg/inch³ (3 s.f.) ✓

This is a strange unit to select as it is a combination of imperial and metric units. It might be useful if you wished to find the weight in kg of something for which you had the measurements in inches. It might be more sensible to use ounces per cubic inch. ✓

Divide by 1000 to convert g to kg.

If you already found this to answer part **a)**, you can refer to the working in part **a)**. Don't just write an answer as examiners don't always get to see every part of the question.

It asks for an approximation as your reading from the graph is unlikely to be completely accurate. By taking the biggest clear point, you can minimise the effect of your error.

5.12 Graphs

You should be able to create a graph given a scale, ratio or proportional relationship.

Example 5.12

The ratio of butter to flour when making pastry is 10 : 22.

Write an equation for y in terms of x to show the relationship between the amount of butter (x) and the amount of flour (y). *(2 marks)*

The ratio is 1 : 2.2 as a unit ratio. ✓

$y = 2.2x$ ✓

By considering the unit ratio, you can see that to go from the value of butter to the amount of flour you would multiply by 2.2

5.13 Rate of Change as a Point on a Graph

The gradient of a graph is the rate of change. If the graph is curved then the rate of change is itself changing, but you can find the instantaneous rate of change by considering the gradient at that point. If the graph is linear then the rate of change is constant.

Example 5.13

A car is travelling along a straight track. The graph shows the velocity against time for the journey.

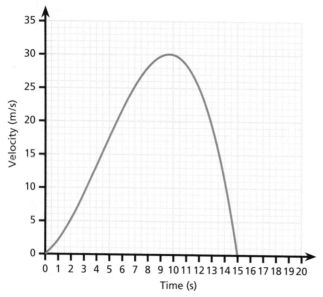

Gradient is 0 at the peak of the curve.

a) Acceleration is the rate of change of velocity against time. At approximately what time is the acceleration of the car 0? *(1 mark)*

The gradient is 0 and therefore the acceleration is 0 at $t \approx 9.7\,s$ ✓

The gradient of the curve is the rate of change of the velocity with regard to time, or acceleration. The acceleration is 0 when the gradient is 0. It is an approximation as it requires you to read the value from the graph. The accurate value is 9.6978... but you could never read a graph that accurately. It clearly falls between 9.5 and 10 though, so a value in this range is acceptable.

b) Estimate the acceleration at $t = 5\,\text{s}$. *(4 marks)*

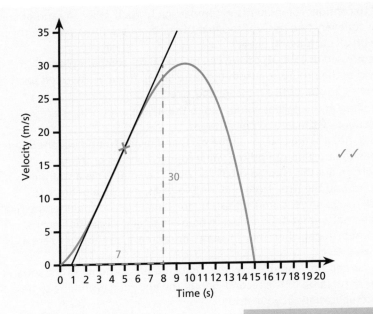

The gradient of the line at $t = 5\,\text{s}$ is:

$a = \dfrac{30}{7} = 4.2857...$ ✓

$\quad = 4.3\,\text{m/s}^2$ (2 s.f.) ✓

Whilst the question does not ask for a suitable degree of accuracy, it is good practice to round your final answer, especially when approximating.

Once again, the approximation is because you need to read a value from the graph. The line you draw and the points you interpret may be different but it should all support your final answer. The accurate answer would be 4.25 but it is not possible to get this accuracy when drawing on the gradient line and then reading values from it.

 For more on the topics covered in this chapter, see pages 16–19 of the Collins Edexcel Maths Higher Revision Guide.

Ratio, Proportion and Rates of Change: Key Notes

- Ratios are used to consider components in comparison to each other. A ratio of 2 : 3 means that for every two parts of the first thing, there are three parts of the second thing, with all parts being equal.
- A unit ratio is expressed as 1 : n. This form can be useful for comparing ratios.
- If you are looking at the proportion from the ratio (2 : 3), there are two red parts (for example) out of a total of five parts, giving $\frac{2}{5}$ red.
- Proportion is looking at the part as compared to the whole; proportions are expressed in fractions, decimals and percentages.
- If you are sharing into a ratio, add up all the parts and divide to find out the value of each part, then multiply up to find the value for each person or thing.
- Ratio and proportion questions could catch you out. They may have a strange context, ask you to find the total from one, or two parts, involve algebra, etc.
- You can do quick checks on your working as you go along (e.g. add together the final ratio to check the total is what it should be) to help avoid 'silly' calculation errors.
- Quantities are proportional when they increase, or decrease, at the same rate as each other. $y \propto x \rightarrow y = kx$
- To make things harder, one **variable** may end up being proportional to the square, root or some other function of the second. Set up the equation each time to help you take these complications in your stride.
- Inverse proportion is when one quantity increases and the other decreases proportionally (i.e. if one doubles, the other halves). $y \propto \frac{1}{x} \rightarrow y = \frac{k}{x}$
- Interest is a proportional increase. In **compound interest**, the same interest is applied repeatedly (note that the second time the interest is applied, it is to the original amount and the interest from the first 'year'). For example, £100 with 10% per annum compound interest ('per annum' means 'every year'):
  After 1st year £110   After 2nd year £121   After 3rd year £133.10
- Using decimal multipliers is an efficient way to calculate repeated increases (interest/growth) and decreases (depreciation/decay). To increase by 12% would be a multiplier of 1.12. To decrease by 7.6% would be a multiplier of 0.924
- In proportionality, the use of algebra and graphs will take questions to a higher level. You need to be able to recognise these graphs:

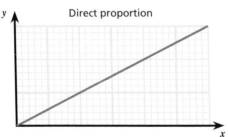

Direct proportion

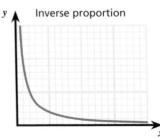

Inverse proportion

Geometry and Measures

Geometry has been a cornerstone of mathematics for thousands of years. The Ancient Greeks used geometrical proofs, where today you might choose to use algebra. The Egyptians used certain properties in building the pyramids, for example the properties of Pythagorean triples in order to create right angles.

6.1 Constructions and Loci

Loci are a set of points that obey a particular rule. You are often required to use construction methods to represent them. Remember that whenever you complete a construction you need to leave in all your construction marks; they should be faint but visible so it is clear that you have followed the correct technique.

Example 6.1

Bill has a goat in a rectangular field, ABCD. He ties it to point H on a rectangular shed with a rope that is 2.2 m long, attached 1.5 m above the horizontal ground.
The goat can reach the ground 30 cm further than the length of the rope.

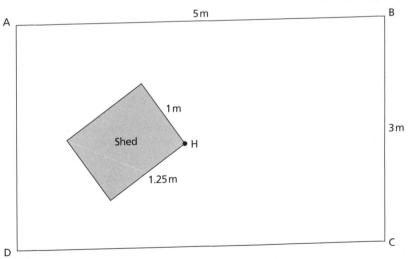

Scale: 2 cm = 1 m

Geometry and Measures

a) What is the furthest horizontal distance that the goat can reach from point H?
(2 marks)

The goat can reach 30 cm + 2.2 m = 2.5 m from the point on the shed to the end of the goat's nose.

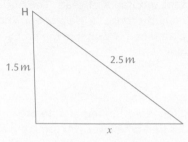

$$2.5^2 = \frac{25 \times 25}{100}$$

$$= \frac{625}{100}$$

$$= 6.25$$

$$
\begin{array}{r}
25 \\
\times\ 25 \\
\hline
12\overset{1}{5} \\
500 \\
\hline
625
\end{array}
$$

Using Pythagoras' theorem

$x^2 = 2.5^2 - 1.5^2$ ✓

$x^2 = 4$

$x = 2\,m$

$$1.5^2 = \frac{15 \times 15}{100}$$

$$= \frac{225}{100}$$

$$= 2.25$$

$$6.25 - 2.25 = 4$$

$$
\begin{array}{r}
15 \\
\times\ 15 \\
\hline
7\overset{2}{5} \\
150 \\
\hline
225
\end{array}
$$

The goat can reach 2 m along the ground from the shed. ✓

Bill wants to plant some trees along the perimeter of the field.
Each tree must be planted within 0.5 m of the perimeter fence.
Bill wants the trees to be closer to the fence AB than to BC.
If the goat can reach the trees when they are little she will eat them, so Bill will plant the trees out of the goat's reach.

b) Shade the region where Bill should plant his trees. *(7 marks)*

There are three separate constructions to consider here.

Constructing a line 0.5 m inside the fence: Measure 1 cm (represents 0.5 m) along each of the edges of the field and make a small mark, then join up faintly. This gives you a smaller rectangle in the middle with some squares in each corner. Still using your ruler, go over the smaller rectangle.

Constructing an angle bisector of ABC: Construct two faint **arcs** an equal distance from B on BA and BC. Re-centre your compasses on each arc where it crosses the line (at each intersection). Draw a faint arc roughly where you expect the line to go, equal distance from each intersection. To do this, it is important to keep the compasses set the same. These should produce a cross and you can then use your ruler to draw the line through B and the cross all the way across the field.

Constructing the region reachable by the goat: The goat can reach 2 m along the ground. Faintly extend the lines of the shed away from H. Set your compasses to 4 cm (2 m). Draw an arc that goes from one faint line to the other. When the goat reaches the corner of the shed, the rope is effectively shorter. You can calculate the new length, or set your compasses by placing the point on the corner of the shed and matching the pencil to the arc where it meets the faint line. Use this to draw the arc until it meets the shed again.

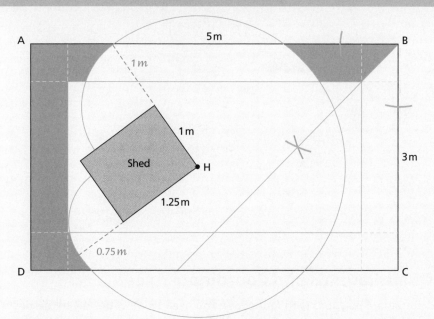

Angle bisector ABC. ✓ ✓   Small arcs with radius 1 m and 0.75 m
Rectangle construction. ✓   respectively. ✓ ✓
Large arc with radius 2 m. ✓   Correct area shaded that meets all criteria. ✓

6.2 Angle Properties and Reasoning

You are expected to be able to reason your answers when it comes to angle properties. Short sentences that have the name of the property and the logic behind it, alongside clear steps of working out, will show you understand and can communicate the understanding clearly.

Example 6.2 📱

Find the values of x and y. State clearly the reasoning for each stage of your calculation.   *(3 marks)*

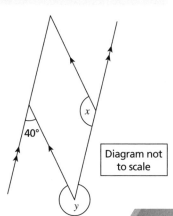

Diagram not to scale

Geometry and Measures

Without the reasoning explained, you will not get the marks for this question.

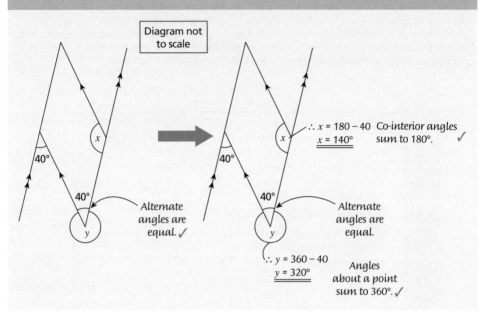

Diagram not to scale

∴ x = 180 − 40   Co-interior angles
x = 140°   sum to 180°. ✓

Alternate angles are equal. ✓

∴ y = 360 − 40
y = 320°   Angles about a point sum to 360°. ✓

Alternate angles are equal.

6.3 Quadrilaterals and Coordinate Axes

Quadrilaterals are **polygons** with four edges. You need to know the key properties of these quadrilaterals: square, rectangle, parallelogram, trapezium, kite and rhombus. Coordinate axes are a way of describing two-dimensional space.

Example 6.3 🖩

Evren has drawn some quadrilaterals.

He wants to plot them on coordinate axes. He has started finding the coordinates of one of the quadrilaterals: A (–3, 1); B (6, 3); D (–4, –3). C lies on the y-axis.

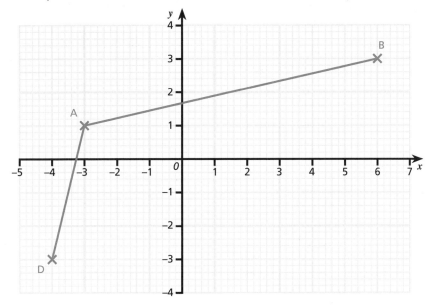

Find the name and dimensions of the quadrilateral. Justify each stage of your answer.
(8 marks)

Angle DAB is greater than 90° so it isn't a rectangle.

If it was a parallelogram then C would be 4 down and 1 left from B = (5, –1), which isn't on the y-axis, so it must be a <u>kite</u>. ✓ ✓

C (0, y)

As AD ≠ AB

AD = CD

AB = BC

To find the length of a line

$$\sqrt{(x_1 - x_2)^2 + (y_1 - y_2)^2}$$

As is often the case, there are many ways of doing this question. You could, for example, use a construction method to find the missing vertex. This answer takes an algebraic approach but, as long as you justify your working and show your method clearly, it is no better or worse than any other method. A possible construction method is shown at the end of the answer.

When using this equation, it doesn't matter which way round you decide to use the x value for A and the x value for B, but it should be consistent with which way round you use the y values.

AD = CD

$$\sqrt{(x_A - x_D)^2 + (y_A - y_D)^2} = \sqrt{(x_C - x_D)^2 + (y_C - y_D)^2}$$

$$\sqrt{(-3 - -4)^2 + (1 - -3)^2} = \sqrt{(0 - -4)^2 + (y - -3)^2}$$

$$(-3 - -4)^2 + (1 - -3)^2 = (0 - -4)^2 + (y - -3)^2$$

$$1^2 + 4^2 = 4^2 + (y - -3)^2$$

$$1^2 = (y - -3)^2$$

$$y + 3 = \pm 1$$

$$\underline{y = -2 \text{ or } y = -4} \checkmark$$

It would be very easy here to lose an answer. If it is a quadratic that you are solving, there will be two answers; if you only get one you should check your working. (It is possible to have no answers but if you have found one then there should be a second one too.)

AB = BC

$$\sqrt{(x_A - x_B)^2 + (y_A - y_B)^2} = \sqrt{(x_C - x_B)^2 + (y_C - y_B)^2}$$

$$\sqrt{(-3 - 6)^2 + (1 - 3)^2} = \sqrt{(0 - 6)^2 + (y - 3)^2}$$

$$(-9)^2 + (-2)^2 = (-6)^2 + (y - 3)^2$$

$$85 = 36 + (y - 3)^2$$

$$49 = (y - 3)^2$$

$$y - 3 = \pm 7$$

$$\underline{y = 10 \text{ or } -4} \checkmark$$

Compare with answers above; $y = -4$ as it is true for both cases. ✓

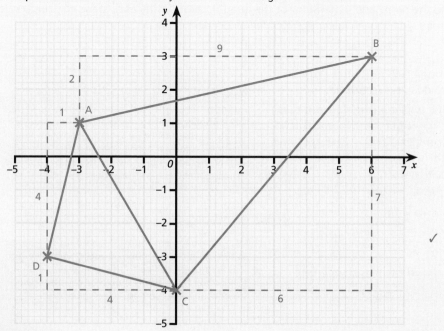

It is a kite

$AB = BC = \sqrt{9^2 + 2^2} = \sqrt{85}$ ✓

$AD = CD = \sqrt{1^2 + 4^2} = \sqrt{17}$ ✓

Alternative method

Having two different ways of finding your answer can be useful and here is an alternative method. If questions don't specify a method, you can choose, but don't spend too much time deliberating. Try one method and, if it doesn't seem to be working out, try something else.

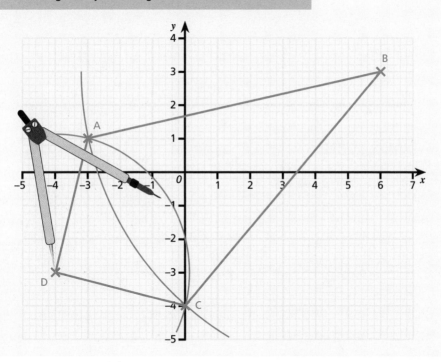

6.4 Congruent Triangles

Congruent shapes are exactly the same as each other. That means their corresponding dimensions and angles match.

Example 6.4

a) Determine if triangles B, C, D, E, F and G are congruent or not congruent (or if there is not enough information to know) with triangle A. Justify your answers. *(4 marks)*

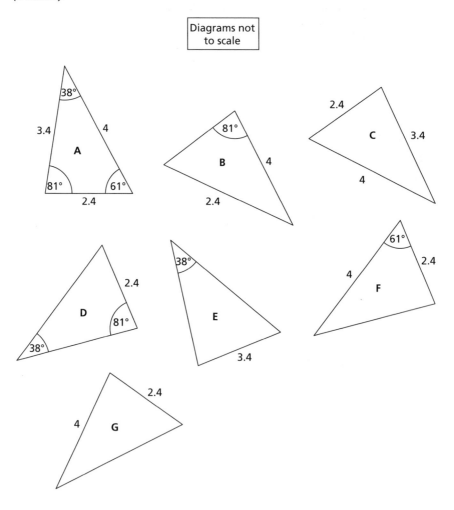

Diagrams not to scale

Triangle B is not congruent as the side opposite the 81° would need to be 4, not 2.4.

Triangle C is congruent (SSS).

Triangle D is congruent (AAS). ✓

Triangle E is not congruent since the opposite side to 38° would need to be 2.4, which it is not. ✓

Triangle F is congruent (SAS). ✓

Triangle G: it is impossible to tell with only two pieces of information. ✓

One mark would be lost for each incorrect answer or missing justification.

b) Explain why having three matching angles is not enough to prove the congruence of two triangles. *(1 mark)*

For congruence, all the corresponding sides need to be the same lengths. If you only have the angles, the triangles could be similar but effectively an enlargement, which is not congruent. ✓

c) Explain why having two matching angles and one matching side is enough to prove congruence in two triangles. *(1 mark)*

If you know two angles in a triangle, it is the same as knowing three since they add up to 180°. If you have ASA, then you know it is similar (matching angles) and the length of the side tells you that the scale factor is 1, so they are congruent. ✓

It is important to note that the side must be corresponding in the same place relative to the given angles, in order to make this comparison.

6.5 Similarity

Similar shapes are those that have the same angles, in the same order. The shapes can be bigger or smaller than each other. The ratio of the sides remains the same. **Trigonometry** is based on the idea that if you know two angles (a right angle and another) in a triangle then the ratio between the sides is the same, however big the triangle.

Example 6.5

The diagram shows three triangles: ABC, AB'C' and ADE.

AC = 40 mm and AD = 8 cm
BC is parallel to DE.

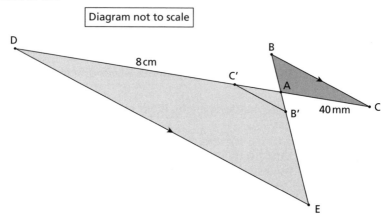

Diagram not to scale

D

8 cm

C'

B

A

B'   40 mm   C

E

a) Prove that triangle ABC and ADE are similar. *(3 marks)*

Angle EAD = Angle BAC
(vertically opposite angles
are equal)

Angle ADE = Angle BCA
(alternate angles are
equal) ✓

Angle AED = Angle ABC
(alternate angles are
equal) ✓

This proves the triangles
are similar as all the
angles are the same. They
are not congruent because
the corresponding sides AC
and AD are not equal. ✓

Finding that each angle in one triangle is equal
to each angle in the second triangle is enough to
prove similarity. If you have a polygon with more
than three angles, you also need to check that
the angles appear in the same order. For example,
these two quadrilaterals have the same angles but
are not similar!

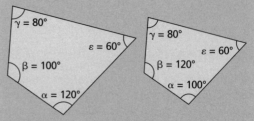

Can you explain why it is enough in a triangle but
not a quadrilateral?

b) Triangle ABC is enlarged by scale factor −0.5 to create triangle AB'C'. What proportion of ADE is left uncovered by AB'C'? *(4 marks)*

An enlargement of ABC is similar to ABC, so is similar to ADE.

40 mm = 4 cm

AC' = AC × 0.5 (the negative can be ignored as it gives direction from A but doesn't affect the magnitude)

4 × 0.5 = 2

AC' = 2 ✓

4AC' = AD

Scale factor from AB'C' to ADE is 4.

1 : 4 ✓

Ratio of areas of the two shapes

$$1^2 : 4^2$$

$$1 : 16$$

So AB'C' is $\frac{1}{16}$ of ADE.

So $\frac{15}{16}$ of ADE is left uncovered. ✓ ✓

> Magnitude is the size of the length. You know from the diagram which direction it is going in, so the calculation is slightly simplified by ignoring the negative. It is a way of translating between **scalar** and vector quantities.

> You could find the scale factor for ADE from ABC, then combine scale factors. Finding the unknown edge AC', then finding the scale factor, is the method used in the example.

6.6 Transformations (including Negative and Fractional Scale Factors)

Transformations include **rotations, reflections, translations** and **enlargements**. Marks are often lost by certain elements of the question being ignored. Centres of rotation and lines of symmetry often require you to have a sound understanding of coordinates and equations of straight lines.

Example 6.6 🔲

A hexagon, A, is shown on the axes.

a) Enlarge A with a **centre of enlargement** (4, 6) and a scale factor of −1. Label your new shape, B. *(2 marks)*

b) Reflect B in the line $y = 3$ and label the new shape, C. *(1 mark)*

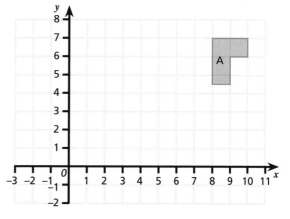

c) Rotate C 180° about (4, 1). Label your new shape, D. *(2 marks)*

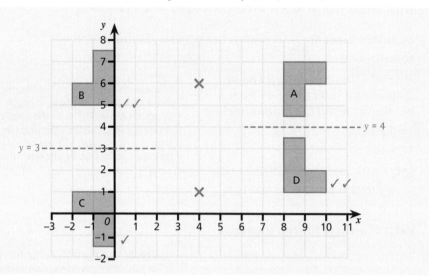

d) Describe fully the single transformation that maps D on to A. *(1 mark)*

Reflection in the line $y = 4$ ✓

Describe fully the single transformation. Marks are often lost by students missing part of the description or describing it in more than one step, typically a reflection or rotation, then a translation.

6.7 Perimeter of Compound Shapes and the Cosine Rule

Perimeter is the distance around the outside of a shape. In practical terms, it might be needed to consider fencing or used to help find the surface area of a prism. It can be found by adding edges. The perimeter of a circle has a special name, circumference, and is found using the formula πd or $2\pi r$.

The cosine rule is used in non-right-angled triangles when you know three out of four key pieces of information and you need to find the fourth, unknown, quantity from: the three sides (a, b, c) and an angle (A).

Example 6.7

Sabiha is making scenery for a school play.
A shape is formed of rectangles, a triangle and a semicircle.

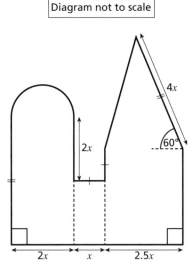

Diagram not to scale

Before painting the piece of scenery, Sabiha sticks tape along the edges to protect from splinters. She cuts four pieces of tape, which allow a small overlap between each piece. Each piece is 3 m long.

Use this information to make an inequality with x as the subject. *(8 marks)*

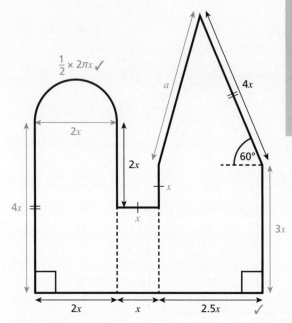

$\frac{1}{2} \times 2\pi x$ ✓

a

$4x$

$60°$

$2x$

x

$4x$

$2x$

x

$3x$

$2x$

x

$2.5x$ ✓

Annotating your diagram is really useful. It shows what you have found and what you need to find. To find the unknown lengths, consider the line markings that show which lines are of equal length. To find the length of the right-hand vertical $4x - 2x + x = 3x$.

a, use cosine rule:

$a^2 = b^2 + c^2 - 2bc \cos A$ ✓

$a^2 = (4x)^2 + (2.5x)^2 - 2 \times 4x \times 2.5x \times \cos 60$

$\quad = 16x^2 + 6.25x^2 - 20x \times \frac{1}{2}$

$\quad = 12.25x^2$

$a = \sqrt{12.25x^2}$ ✓ (Ignore negative root as we are

$\quad = 3.5x$ ✓ looking for a positive length.)

$\sqrt{12.25}$ might seem like a complicated calculation without a calculator. You know that it must end in .5 for the decimal to be .25. You also know that it is between 3 and 4 ($\sqrt{9}$ and $\sqrt{16}$) so the answer is 3.5. Check by squaring it. $3.5 \times 3.5 = 12.25$

Perimeter = $4x + \pi x + 2x + x +$
$\qquad x + 3.5x + 4x + 3x +$
$\qquad 2.5x + x + 2x$

$\qquad = (24 + \pi)x$ ✓

$(24 + \pi)x < 4 \times 3$

$(24 + \pi)x < 12$

$\qquad x < \frac{12}{24 + \pi}$ ✓ ✓

When the perimeter has multiple sections, use a methodical approach to make sure you don't miss any out. In this example, it starts at one vertex and works around each edge going clockwise.

You know to use < ('less than') as the question says there is some overlap of the tape. With no overlap the perimeter would be 12 m, so it must be less than 12 m.

6.8 Circle as a Graph Area

The equation of a circle, centred at the origin, is $x^2 + y^2 = r^2$, where r is the radius.

Example 6.8

Part of the graph $x^2 + y^2 = 100$ is shown.

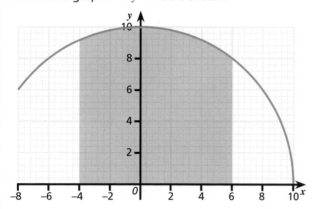

You should recognise the equation as that of a circle.

a) By using five fixed width intervals, estimate the shaded area. *(6 marks)*

First split the area into the five widths. As you know it is a circle and symmetrical about the y-axis, it is possible to only have to calculate three different areas before combining them.

In the first instance, the method is to find the height at the midpoint then calculate the area of the rectangles. In the alternative method, the areas will be approximated to trapezia.

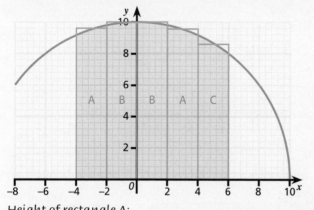

Height of rectangle A:

$3^2 + y^2 = 100$

$y^2 = 100 - 9 = 91$

$y = \sqrt{91}$

At this point, you could turn this into a number but, as the **square root** is accurate and the decimal version would be truncated, it is worth leaving it in this form for now.

Area of A = $2\sqrt{91}$ ✓
So area of both As = $4\sqrt{91}$

Height of rectangle B:
$1^2 + y^2 = 100$
$\quad y^2 = 100 - 1 = 99$
$\quad\quad y = \sqrt{99}$
Area of B = $2\sqrt{99}$ ✓
So area of both Bs = $4\sqrt{99}$

Height of rectangle C:
$5^2 + y^2 = 100$
$\quad y^2 = 100 - 25 = 75$
$\quad\quad y = \sqrt{75}$
Area of C = $2\sqrt{75}$ ✓

Total area ≈ $4\sqrt{91} + 4\sqrt{99} + 2\sqrt{75}$ ✓
≈ 95.2775736...
≈ 95.3 (3 s.f.) ✓

> There are no units associated with this question but, if there were, it would be an area so the units would be squared.

Alternative method

x	−4	−2	0	2	4	6	
y	$\sqrt{84}$	$\sqrt{96}$	10	$\sqrt{96}$	$\sqrt{84}$	8	✓ ✓
	a_0	a_1	a_2	a_3	a_4	a_5	

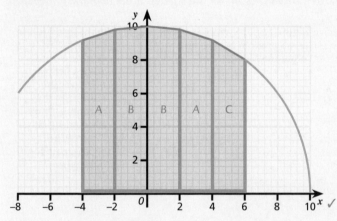

Area of a trapezium = $\frac{a+b}{2}h$

Total area using the trapezium rule (adding the areas of all the trapezia together)

≈ $\frac{1}{2}h(a_0 + 2(a_1 + a_2 + a_3 + a_4) + a_5)$ ✓

≈ $\frac{1}{2} \times 2 \times (\sqrt{84} + 2(\sqrt{96} + 10 + \sqrt{96} + \sqrt{84}) + 8)$ ✓

≈ 94.68729005...

≈ 94.7 (3 s.f.) ✓

> Since this is only an estimate, the two different methods give slightly different answers, which is expected.

b) Find the exact area of the shaded region. What is the percentage error in your approximation from part **a)**? *(10 marks)*

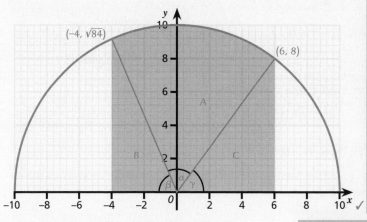

Writing out your thought process and method can help both you and the examiner! Labelling and drawing on your diagram can help. It may be crowded from part **a)** so consider re-sketching it.

By splitting the area into component parts, you get two triangles and a **sector** of a circle.

Area of a sector is the proportion of the circle $\frac{\alpha}{360} \times \pi r^2$. You know the radius is 10.

First you need to find α.

$\tan\beta = \frac{\sqrt{84}}{4}$

$\beta = \tan^{-1}\left(\frac{\sqrt{84}}{4}\right) = 66.42182...$ (B in calc. memory) ✓

$\tan\gamma = \frac{8}{6}$

$\gamma = \tan^{-1}\left(\frac{8}{6}\right) = 53.130102...$ (C in calc. memory) ✓

$\alpha = 180 - 66.42182... - 53.130102...$ (Angles on a straight line sum to 180°)

$= 60.4480761...$ (A in calc. memory) ✓

Area of A $= \frac{60.448076...}{360} \times \pi 10^2 = 52.7508977...$ (D in calc. memory) ✓

Area of B $= \frac{1}{2}bh = \frac{1}{2} \times 4 \times \sqrt{84} = 4\sqrt{21}$ ✓

Area of C $= \frac{1}{2}bh = \frac{1}{2} \times 6 \times 8 = 24$ ✓

Total area $= 24 + 4\sqrt{21} + 52.7508977...$

$= 95.08120052...$ (E in calc. memory) ✓

Use your calculator's memory to store this answer accurately. Making a note to yourself means you can then use it in later parts of the calculation confidently.

Having got to this point, it is tempting to round this as your final answer. But checking the question shows that you are expected to find the percentage error in YOUR approximation. If you didn't manage to find an approximation, it is worth making one up for the marks in this second part.

Percentage error = $\dfrac{\text{Real value} - \text{Approximate value}}{\text{Real value}} \times 100$

$= \dfrac{95.0812... - 95.277...}{95.0812...} \times 100 = -0.206531$ ✓

The approximation is 0.21% (2 d.p.) over-estimated. ✓

Note to self: calculation is

$= \dfrac{E - F}{E} \times 100$

where

$F = 4\sqrt{91} + 4\sqrt{99} + 2\sqrt{75}$

(From part **a**)

> Found using the approximation from the first method.
> If you approximated using trapezia, then you should have
> an underestimate of 0.41% (2 d.p.).

6.9 Circle Theorems and Polygons

The circle theorems you have learned can be applied in a number of ways to solve geometrical problems.

Example 6.9 🖩

A **regular polygon** intersects a circle, of centre O.
BL is a **tangent** to the circle.
AO is a straight line.

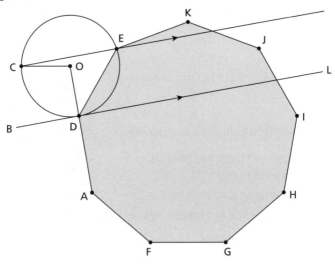

Find angle COD. Justify each stage of your working. *(5 marks)*

Angle COD is double the angle DEC.

ADEKJIHGF is a nonagon, it has nine sides.

Each interior angle = $\frac{180 \times 7}{9}$ = 140° ✓

∠ODE = 180 − 140 = 40° (angles on a straight line sum to 180°)

Using the formula for **interior angles** of a polygon $(n - 2) \times \frac{180}{n}$.

You could also do $\frac{360}{n}$ to find the exterior angle, then subtract from 180 (angles on a straight line).

If you find something that you decide isn't actually useful, don't cross it out. It might be useful later. If you make a mistake, also don't cross it out until you have something to replace it with.

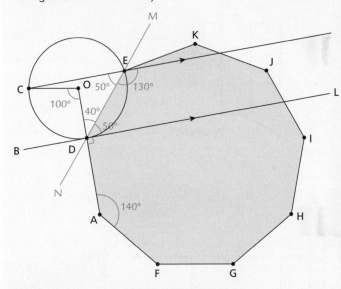

Annotating the diagram helps. Consider extending lines in order to get more information; in this case it makes **corresponding** and **alternate angles** more apparent.

∠LDA = 90° (radius of circle meets a tangent to the circle at a right angle) ✓

∠LDM = 140 − 90 = 50° (∠ADE is the interior angle of the nonagon) ✓

∠CED = 50° (alternate angles are equal) ✓

∠COD = 100° (angle at the centre of the circle is double that at the edge when subtended by the same arc) ✓

You may find extra information that you decide you don't actually need (for example, here 130° is labelled at E; co-interior angles then angles on a straight line to get the 50°). There isn't a right or wrong way to do it, but make sure your working covers all the steps needed to find the answer.

6.10 Plans, Elevations and Projections

Plans, elevations and projections are used in many areas. A key example would be by architects, engineers and builders to communicate their ideas, both to each other and to their clients. Being able to understand and visualise shapes from plans and elevations is a very useful skill.

Example 6.10

Alison is designing a new office block. Below are the plan and two elevation drawings of her design.

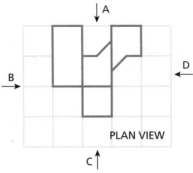

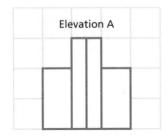

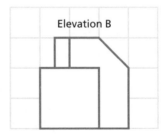

a) In the space provided, complete the remaining elevations. *(4 marks)*

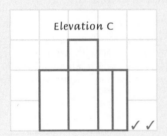

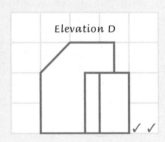

The building has three main parts with each being fully constructed before the next is started. Alison visits the site when construction of the central tower has been completed but neither of the wings has yet been started. The central tower is shown shaded on the plan.

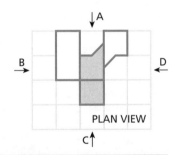

PLAN VIEW

b) In the space below, draw a projection of the building at this stage of construction. *(2 marks)*

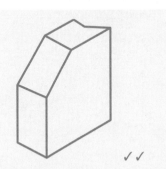

✓ ✓

To draw your projection, choose a corner to start from. This is one of four possible projections, looking from corner CD. Make sure the lines that are meant to be parallel are parallel, so the flat roof and the ground for example. If the proportions aren't quite how you want them, don't worry as long as you are able to represent the shape. Don't waste too much time on small details!

Alternative answers:

6.11 Bearings, Measurement and Scales

Bearings are used to communicate about direction. Generally a bearing is given from point A to point B, or of point B from point A. Make sure you start at the correct one. Bearings are a measure of turn (number of degrees) and are taken from the north. You might be asked to create a construction or to use given bearings to find an answer using angle facts. Look out for parallel lines (the north lines). Questions can require a geometric approach to solve them but can also include scales and ratios.

Example 6.11

Mia is flying her plane at a constant height. She passes over Bilston and takes a bearing of 060° for 10 km until she is over Chirley. She then flies on a bearing of 146° for 11 km to reach Deeston.

a) Using a suitable scale, construct a diagram to represent Mia's flight and use it to estimate the bearing and horizontal distance she would take in order to be back above Bilston. *(6 marks)*

Consider starting with a little sketch so you can decide where in the space you should begin. Shortening names and details in your workings is fine, especially when they will be obvious. If you choose something less obvious, include a key.

'Suitable scale'

1 cm = 2 km ✓

A suitable scale is one that is as large as possible for the space (to minimise the effect of residual errors) but is also one that is easy to work with.

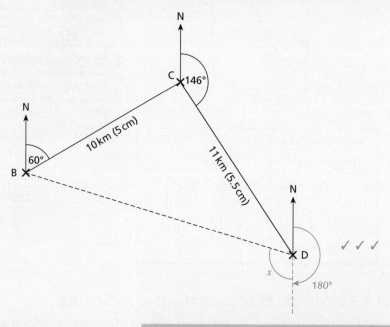

$x \approx 105°$ (based on measurement to the nearest degree)

Bearing of B from D ≈ 285° ✓

BD ≈ 7.7 cm

BD ≈ 7.7 × 2 ≈ 15.4 km ✓

When constructing, a certain error is expected as measurements can only be performed to a certain accuracy. The accurate answer is 285.5...° and the length is 15.37356...km but it is impossible to achieve this accuracy without calculation. Your answer must be right for your diagram and full marks will be achieved if all aspects of your diagram fall within the allowed level of accuracy (generally within a few degrees).

b) Mia is flying at a constant height of 0.5 km. On the return leg of her flight, what is the closest she will be to her friend, who is sitting in a café in the centre of Chirley? *(6 marks)*

Closest when line from Mia to her friend in Chirley is at right angles to her line of flight.

First find y

$d = 180 - 105 - 34 = 41°$

$\sin 41 = \dfrac{y}{11}$

$\quad y = 11\sin 41$ ✓

$\qquad = 7.2166...\,km$ ✓

As Mia is 0.5 km up, using Pythagoras:

Distance $= \sqrt{0.5^2 + 7.2166...^2}$ ✓

$\qquad = \sqrt{52.330027...}$

$\qquad = 7.2339496...$

$\qquad = 7.23\,km$ (3 s.f.) ✓

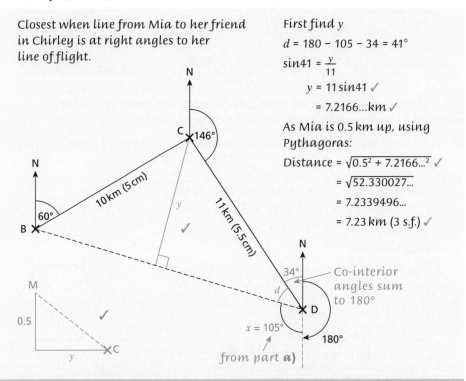

The answer here depends on your answer from part **a)**. This means there will be variation allowed in the answers gaining full marks. For this reason, it is even more important to explain and show each step of your working.

6.12 Compound Area, Volume and Use of Formulae

Compound shapes are ones that you need to split into recognisable parts in order to calculate the individual volumes (or areas) and hence find the total volume (or area). It might be that you need to remove a section or add sections together but spotting the parts is key in solving these questions.

Example 6.12

A children's jigsaw is made up of three prisms: a green cylinder, a red flower and an orange, equilateral triangular prism.

The flower is a regular hexagon, with edge length 1.5 cm. From each edge a semicircle forms the petals. Each prism is 1.5 cm deep.

a) What is the volume of the flower prism? Give your answer to 3 significant figures. *(6 marks)*

> Volume of a prism is <u>the area of the face × length</u>.
>
> Area of face = Area of hexagon + Area of 6 semicircles (3 full circles).
>
> Each circle $A = \pi r^2 = \pi\left(\frac{3}{4}\right)^2 = \frac{9\pi}{16}$ ✓ ✓
>
> Area of hexagon:
>
> A regular hexagon is made of 6 equilateral triangles.
>
> Area of a triangle $= \frac{1}{2}ab\sin C = \frac{1}{2} \times 1.5^2\sin 60 = \frac{9\sqrt{3}}{16}$ ✓ ✓
>
> Total area $= 3 \times \frac{9\pi}{16} + 6 \times \frac{9\sqrt{3}}{16}$ ✓
>
> $= 11.147109... \text{ cm}^2$
>
> Volume $= 11.1471... \times 1.5 = 16.7206636... = 16.7 \text{ cm}^3$ (3 s.f.) ✓

b) The volume of each prism is the same. Kamil is making 100 of these complete jigsaws. A 50 ml tube of paint covers an area of 270 cm². Tubes of red paint cost £1, tubes of green paint cost £0.90 and tubes of orange paint cost £1.05.

How much will the paint cost in total? *(10 marks)*

> Plan:
>
> 1. Find the surface area of each shape.
> 2. Find the dimensions from volume.
> 3. Calculate how much paint for each shape.
> 4. Calculate the cost of paint for each shape.
> 5. Add the costs to find the total cost.

When performing a similar calculation a number of times, a table can be a good way of setting out your working.

Volume $\pi r^2 \times 1.5 = 16.720...$ $r^2 = 3.548...$ $r = 1.8836...$ ✓	Volume Not needed as dimensions are already given.	Volume $\frac{1}{2} \times a^2 \times \sin60 \times 1.5$ $= 16.720...$ $a^2 = 25.74314...$ $a = 5.07377...$ ✓
Surface area $= 2\pi r^2 + 2\pi r \times 1.5$ $= 40.04744558...$ ✓	Surface area $= 2 \times (11.147109...) + 2\pi \times 0.75 \times 3 \times 1.5$ $= 43.499968...$ ✓	Surface area $= 3 \times a \times 1.5 + 2 \times \frac{1}{2}a^2\sin60$ $= 45.12618...$ ✓
Tubes of paint $= 4004.744558... \div 270$ $= 14.832...$ 15 tubes of green paint needed ✓	Tubes of paint $= 4349.9968... \div 270$ $= 16.11...$ 17 tubes of red paint needed ✓	Tubes of paint $= 4512.618... \div 270$ $= 16.134...$ 17 tubes of orange paint needed ✓
Cost of paint $= 15 \times 0.90 = £13.50$	Cost of paint $= £17$	Cost of paint $= 17 \times 1.05 = £17.85$

Total cost of paint = 13.50 + 17 + 17.85 = £48.35 ✓ ✓

6.13 Known Trigonometric Ratios

You need to know the trigonometric ratios (the value of $\sin\theta$, $\cos\theta$ and $\tan\theta$) for $\theta = 0°$, 30°, 45°, 60°, 90°. The ratio tells you the relationship between two of the sides in a right-angled triangle.

Example 6.13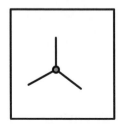

A mast, of height h m, is held in a vertical position by three ropes, which are attached to the horizontal ground near the base of the mast.
Ropes A and B meet the ground at angles of $\theta°$ and $2\theta°$ respectively.

a) In terms of h and θ, find expressions for the lengths of rope A and rope B. *(2 marks)*

Rope A

$\sin\theta = \dfrac{h}{x}$

$x = \dfrac{h}{\sin\theta}$ ✓

Including sketches both shows the examiner what you are doing and can help you see the way to get an answer.

Rope B

$y = \dfrac{h}{\sin2\theta}$ ✓

Less working is shown here as the relationship is the same as for rope A.

Rope C meets the ground h m from the base of the mast.

b) Find an expression in terms of h for the length of rope C. *(1 mark)*

$\sin45 = \dfrac{h}{z}$

$z = \dfrac{h}{\sin45} = h\sqrt{2}$ ✓

From your sketch you should recognise that rope C forms an isosceles triangle. As it is a right-angled triangle, each of the 'base' angles is 45°.
You could also get this result using Pythagoras' theorem,
$z^2 = h^2 + h^2 = 2h^2$
$z = \sqrt{2h^2} = h\sqrt{2}$

The longest rope is 12 m. The mast is going to be raised a further 2 m from its original height of 6 m. The points for attaching the ropes will stay in the same place. Ffion will reuse a rope if it is long enough. Each rope must be one continuous length and new rope comes in an integer value of metres.

c) What additional lengths of rope will Ffion need? *(9 marks)*

Original lengths of rope:

When $h = 6\,m$ the longest rope is either A or C.

Length of C is $6\sqrt{2}\,m \neq 12$ ✓

So rope A must be 12 m long. ✓

$$12 = \frac{6}{\sin\theta}$$

$$\sin\theta = \frac{6}{12} = \frac{1}{2}$$

$$\theta = 30° ✓$$

Rope B:

$$y = \frac{h}{\sin 2\theta} = \frac{6}{\sin 60}$$

$$= \frac{6}{\sqrt{3}/2} = \frac{12}{\sqrt{3}} = \frac{12\sqrt{3}}{3}$$

Length of B = $4\sqrt{3}\,m$ ✓

New rope lengths:

Distance of anchor point from mast:

For rope C it is still 6 m.

For rope A it is $\frac{6}{\tan 30} = \frac{6}{\sqrt{3}/3} = 6\sqrt{3}\,m$

For rope B it is $\frac{6}{\tan 60} = \frac{6}{\sqrt{3}} = 2\sqrt{3}\,m$ ✓

Using Pythagoras with new height to find the new lengths of rope:

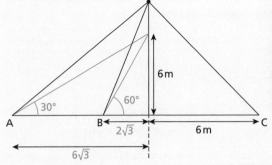

A: $8^2 + (6\sqrt{3})^2 = A^2$

$\quad A = \sqrt{64 + 108} = \sqrt{172}$

B: $8^2 + (2\sqrt{3})^2 = B^2$

$\quad B = \sqrt{64 + 12} = \sqrt{76}$

This is an unusually large question and it is easy to lose yourself in a calculation and get misdirected. Having a brief overview/plan can help keep you on track.

Plan:
1. Find original lengths of rope.
2. Find new lengths of rope.
3. Compare.
4. How much new rope is needed?

These steps are really important to show you haven't just made an assumption in your working. However, it isn't unreasonable to expect a certain value for θ given that this question is non-calculator. But this should be a check in your working rather than the basis for it.

You can use underlining effectively to pick out key bits of information that you might use later. It also signposts your working for the examiner.

Diagram – this could be represented as three separate diagrams, which might make it easier to see what is going on, but it is your working so represent it in the way that makes sense to you. The ropes wouldn't appear like this from any perspective but it works to model the situation.

C: $8^2 + 6^2 = C^2$

$C = \sqrt{64 + 36} = \sqrt{100} = 10$ ✓

Compare:

In each case, the new length is greater than the old length.

Old lengths are:

12, $4\sqrt{3}$, $6\sqrt{2}$

Make fully into surds for comparison:

$\sqrt{144}$, $\sqrt{48}$, $\sqrt{72}$ ✓

New lengths are:

$\sqrt{172}$, $\sqrt{76}$, $\sqrt{100}$

The old rope A can be used as the new rope C.

The other ropes cannot be reused.

New rope?

$8 < \sqrt{76} < 9$ so <u>9 m length of rope needed</u> ✓

$13 < \sqrt{172} < 14$ so <u>14 m length of rope needed</u> ✓

6.14 Vectors – Diagrams, Arguments and Proof

Vectors are used to represent things that have both magnitude and direction. They can be represented by a directed line; the magnitude is represented by the length of the line and the arrow shows the direction. Vectors can be drawn, referred to as column vectors, given names (lowercase letters underlined when handwritten; shown in bold when printed) and manipulated algebraically.

Example 6.14

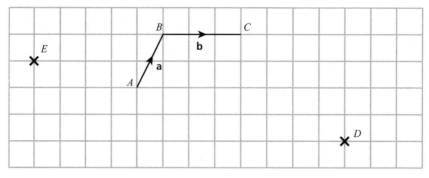

Vector $\overrightarrow{AB}$ = **a** and vector $\overrightarrow{BC}$ = **b**.

a) Find an expression in terms of **a** and **b** to describe $\overrightarrow{AD}$. *(1 mark)*

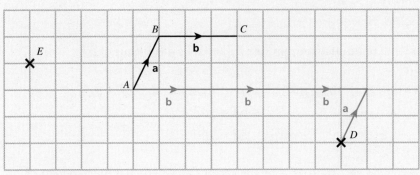

$\overrightarrow{AD} = 3\mathbf{b} - \mathbf{a}$ ✓

b) Find an expression in terms of **a** and **b** to describe $\overrightarrow{DE}$. *(2 marks)*

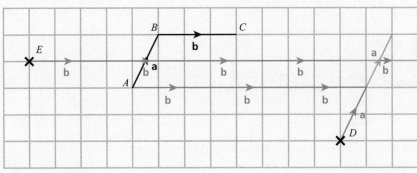

$\overrightarrow{DE} = \frac{3}{2}\mathbf{a} - \frac{9}{2}\mathbf{b}$ ✓ ✓

c) Hence or otherwise, prove that A lies one-third of the way along the line $\overrightarrow{ED}$.
(2 marks)

$$\overrightarrow{DA} = \mathbf{a} - 3\mathbf{b}$$

For EAD to lie on a straight line $\overrightarrow{DE}$ must be a multiple of $\overrightarrow{DA}$:

$$\overrightarrow{DE} = \lambda(\mathbf{a} - 3\mathbf{b})$$

$\frac{3}{2}\mathbf{a} - \frac{9}{2}\mathbf{b} = \frac{3}{2}(\mathbf{a} - 3\mathbf{b})$ so they lie on a straight line and $\overrightarrow{DE} = \frac{3}{2}\overrightarrow{DA}$:

$$\overrightarrow{DA} = \frac{2}{3}\overrightarrow{DE} \checkmark$$

$$\overrightarrow{EA} = \frac{1}{3}\overrightarrow{ED} \checkmark$$

d) Use column vectors to describe the vector $\overrightarrow{CM}$, where M is the **midpoint** of the line BD. *(3 marks)*

$$\overrightarrow{BD} = \begin{pmatrix} 7 \\ -4 \end{pmatrix}$$

$$\overrightarrow{BM} = \begin{pmatrix} 3.5 \\ -2 \end{pmatrix} \checkmark$$

$$\overrightarrow{CB} = \begin{pmatrix} -3 \\ 0 \end{pmatrix}$$

$$\overrightarrow{CM} = \overrightarrow{CB} + \overrightarrow{BM} = \begin{pmatrix} -3 \\ 0 \end{pmatrix} + \begin{pmatrix} 3.5 \\ -2 \end{pmatrix} = \begin{pmatrix} 0.5 \\ -2 \end{pmatrix}$$

$\checkmark$ $\checkmark$

6.15 Compound Shapes, Trigonometric Equations and Algebra

Questions will often expect you to identify which methods to use. In this case, there are a number of different skills being pulled together for you to identify and deal with.

Example 6.15

A regular octagon, of edge length a, is intersected by an isosceles triangle, with two sides of $2a$. The vertex of the triangle is at the centre of the octagon, as shown in the diagram.

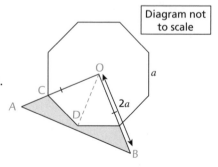

Diagram not to scale

Adding labelling to the diagram will help you to communicate effectively. In this case A, B, C and D have been added and the line OD.

a) Find the exact area of the shaded region in terms of a. *(6 marks)*

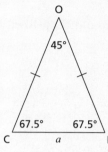

Consider the triangle COD.

Angle at centre = $360 \div 8 = 45°$ (angles about a point sum to $360°$)

Base angles = $\frac{180 - 45}{2} = 67.5°$ ✓ (base angles of an isosceles triangle are equal and all angles in a triangle sum to $180°$)

> If you think of something that might be useful, even if you can't apply it yet, it is worth writing down. In this case, you know you want to find the area, so the area rule is a good thing to have pop into your head. Looking at it, you need some more information before applying it. You now know what you need to find next, OC and OD.

Area = $\frac{1}{2}ab\sin C$ (trig. area rule for non-right-angled triangles)

To use this, find the equal edges OC and OD.

> a here is part of the learnt formula. It doesn't, in this case, refer to the edge a in your triangle. Be careful when translating between formula lettering and that used in the question.

Sine rule:

$$\frac{a}{\sin A} = \frac{b}{\sin B}$$

$$\frac{a}{\sin 45} = \frac{b}{\sin 67.5}$$

$$b = \frac{a\sin 67.5}{\sin 45} ✓$$

> You want your answer in terms of a so this is now ready to substitute into the area formula.

Area = $\frac{1}{2}ab\sin C$

$$= \frac{1}{2}\left(\frac{a\sin 67.5}{\sin 45}\right)^2 \sin 45 ✓$$

$$= \frac{a^2(\sin 67.5)^2}{2\sin 45}$$

> If you chose to use a different pair of sides, you would still get the same answer.
> $$= \frac{1}{2}ab\sin C = \frac{1}{2} \times a \times \frac{a\sin 67.5}{\sin 45} \times \sin 67.5$$
> $$= \frac{a^2(\sin 67.5)^2}{2\sin 45}$$

So area of the two triangles

$$= 2 \times \frac{a^2(\sin 67.5)^2}{2\sin 45} ✓$$

$$= \frac{a^2(\sin 67.5)^2}{\sin 45}$$

> AOB is a right-angled triangle. You can find the missing length using Pythagoras. You know, or can find, all the angles but as you want to know the area and you know AO and BO, you can put these straight into the area formula.

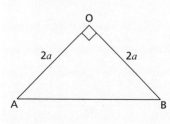

Consider triangle AOB:

Area = $\frac{1}{2}bh$

$$= \frac{1}{2} \times 2a \times 2a = 2a^2 ✓$$

Area of shaded region = $2a^2 - \frac{a^2(\sin 67.5)^2}{\sin 45}$

$$= a^2\left(2 - \frac{(\sin 67.5)^2}{\sin 45}\right) ✓$$

b) Given that the shaded area is 5.0 cm², to 1 decimal place, find the lower bound of the perimeter of the octagon. Give your answer to 1 decimal place. *(5 marks)*

The lower bound of the perimeter will come from the lower bound of the area.
Take area as 4.95 cm² ✓
*Set it equal to the answer from part **a**).*

$a^2\left(2 - \frac{(\sin67.5)^2}{\sin45}\right) = 4.95$

$0.792893... \; a^2 = 4.95$ ✓

$a^2 = \frac{4.95}{0.792893...} = 6.2429591...$

$a = \sqrt{6.2429591...} = 2.49859...$ ✓

Perimeter of the octagon $= 8a$

$= 8 \times 2.49859...$

$= 19.98873152...$ ✓

Lower bound of perimeter $= 20.0$ cm (1 d.p.) ✓

> You can rearrange with the terms in their trigonometric form but as $\left(2 - \frac{(\sin67.5)^2}{\sin45}\right)$ is just a number which you need to divide 4.95 by, you can use your calculator (remember, no rounding until your final answer). The memory function on your calculator can be very useful here.

6.16 3D Trigonometry and Pythagoras' Theorem

One of the ways to make Pythagoras' theorem and trigonometry a bit more challenging is to make you consider them in three dimensions. Generally, the trick to this is breaking it into a series of right-angled triangles.

Example 6.16

A new rule suggests that spaghetti should be limited to 14 cm in length, to help avoid unnecessary scalds when trying to place the pasta in boiling water.
Alfie has a saucepan which is a cylinder. It has a diameter of 12 cm. Alfie fills the saucepan to about three-quarters full of water when cooking pasta. A piece of the new spaghetti will be submerged, but only just, when placed into the pan.

a) What is the full height of the pan? *(3 marks)*

Let y be the full height of the pan.
Using Pythagoras:

$14^2 - 12^2 = \left(\frac{3y}{4}\right)^2$ ✓

$52 = \frac{9y^2}{16}$

$9y^2 = 832$ ✓

$y^2 = 92.444444...$

$y = 9.61480...$

Height of the pan $= 9.6$ cm (1 d.p.) ✓

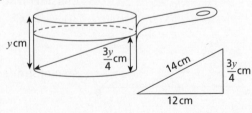

b) Alfie has a container which is a trapezoidal prism, with two planes of symmetry.

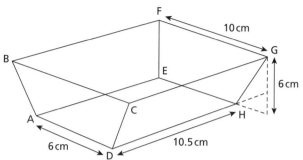

Finding the symmetry of a shape is no longer a question in its own right, but you are still expected to understand descriptions that use symmetry (both rotational and reflection).

Alfie says that the spaghetti will fit well because the diagonal AG is 15 cm. Is he correct? *(5 marks)*

Consider rectangle ACGE. To find AG, you need to find AC.

Use Pythagoras' theorem to find AG.

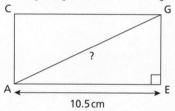

When a situation is complicated, considering the shape 'broken up' can help. Re-draw your diagrams at each stage. It is your working, so include the detail you need. It also helps the examiner to clearly understand what you have done and makes it easier for them to give you all the marks you deserve.

Trapezium ABCD has one line of symmetry through it.

AQ = DP

10 − 6 = 4

4 ÷ 2 = 2

AP = 6 + 2 = 8 cm ✓

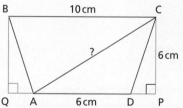

It is possible to go straight for $\sqrt{a^2 + b^2 + c^2}$ when presented with Pythagoras' theorem in 3D. In this case, it is more complicated than a standard cuboid so splitting it into two steps can help to show where the different values have come from.

$AC = \sqrt{6^2 + 8^2}$

$= \sqrt{100}$

$= 10$ cm ✓

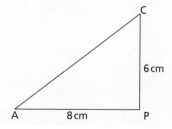

$AG = \sqrt{10.5^2 + 10^2}$

$= \sqrt{110.25 + 100} = \sqrt{210.25}$

$AG = 14.5\,cm$ ✓ ✓

Alfie is right that a piece of spaghetti will fit on the diagonal AG, but he is wrong that AG = 15 cm as it is actually 14.5 cm. ✓

The spaghetti won't fit well because it would need to be at an angle so it wouldn't be a very efficient use of space.

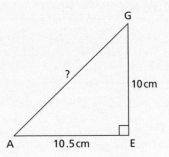

c) Find the angle that the line DG makes with the horizontal. *(5 marks)*

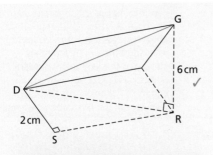

$DR = \sqrt{10.5^2 + 2^2}$

$= \dfrac{\sqrt{457}}{2}$ ✓

$\alpha = \tan^{-1}\left(\dfrac{6}{\sqrt{457}/2}\right)$

$= \tan^{-1}\left(\dfrac{12}{\sqrt{457}}\right)$ ✓

$= 29.3070800...$

$\underline{\alpha = 29.3°\ (3\,s.f.)}$ ✓

Visualising the shapes involved can be hard. Sketching out the face that is needed can help to stop you being distracted by all the other lines, measurements and angles.

For more on the topics covered in this chapter, see pages 20–23, 42–45, 52–55, 74–79 & 102–105 of the Collins Edexcel Maths Higher Revision Guide.

Geometry and Measures: Key Notes

- You need to know the names and properties of polygons (up to 10 sides) and circles. For triangles and quadrilaterals, you need to know the properties (angles, parallel lines, etc.) and the names (isosceles, kite, equilateral, trapezium, etc.). Make sure you know how to spell them correctly.
- A regular polygon has equal edges (and equal angles).
- When constructing, leave in your construction marks – if there aren't any, you might not get credit. Practise working with your pair of compasses for accuracy.
- When working with angle reasoning, you need to <u>justify</u> each stage of your working. Even if your numerical answer is right, you won't get marks unless you can justify and explain your steps.
- There is a lot of assumed knowledge for this topic, including circle theorems, angle rules, trigonometric values and formulae for areas and perimeters/circumference. See Chapter 2 for ideas of how to help learn rules and formulae; see Chapter 10 for what formulae are given and those you need to learn.
- Geometry can get context heavy. Try to see the maths the examiners are testing you on and don't get too hung up on the context. If you need to make assumptions, state them clearly.
- To describe a transformation fully:
  - Reflection: give the line of reflection, e.g. $x = 2$, $y = -4$, $y = x$, etc.
  - Rotation: give the **centre of rotation**, e.g. $(2, -2)$; the amount of turn, e.g. through 90°; and the direction of turn, e.g. clockwise
  - Enlargement: give the centre of enlargement, e.g. $(0, 3)$, and scale factor, e.g. 0.4
- Enlargements can have both fractional and negative scale factors. Make sure you know what to do with them if they come up.
- If a question involves a right-angled triangle, consider:
  - Pythagoras' theorem if you have two sides and need to find the third side.
  - Trigonometry (SOH CAH TOA) if you have two sides and want to find an angle or have an angle (as well as the right angle) and a side but want to find another side. The trigonometric functions tell you what the ratio between two particular sides is in a set of similar triangles.
- If you are stuck, or not sure what the question is asking for, think about what you can do. If there is a rectangle in the diagram, you can find the area, perimeter, diagonals, etc.
- Vectors are quantities with both direction and magnitude (size). They can be expressed as column vectors, e.g. $\begin{pmatrix} 4 \\ -2 \end{pmatrix}$ which means 4 right and –2 up (or 2 down). Column vectors can also describe translations of objects.
- To write a vector in algebraic terms, underline the vector values (in printed text they are shown in bold).

Probability is used to look at the chance of something happening; a particular outcome to an event. It is useful in a variety of ways, from game theory to genetics. You will need to have a good understanding of working with decimals, fractions and percentages to tackle this topic successfully. Questions tend to be given in a context, a 'real life' problem to solve, and can get quite wordy, so refer to Chapter 9 for more help.

7.1 Counters in a Bag – Theoretical Probability

The classic example is of randomly selecting coloured counters (sweets, marbles, coins, etc.) from a bag (pot, jar, etc.). The theoretical probability of an outcome (e.g. blue) can be found by considering the number of blue counters there are and dividing it by how many counters there are altogether. This is a base for many questions but can be made into a higher level question by making you find the number of counters given a probability or by asking you to work in algebraic terms.

Example 7.1

There are three different colours of counters in a bag. There are 4 purple counters, 6 orange counters and 5 yellow counters.

a) Maya takes a counter randomly from the bag. What is the probability she gets a purple counter? *(1 mark)*

There are 4 + 6 + 5 = 15 counters in the bag altogether.

$P(purple) = \frac{4}{15}$ ✓

Probability is:

$$\frac{\text{Number of favourable outcomes}}{\text{Total number of possible outcomes}}$$

b) Maya replaces her counter, then adds more purple counters to the bag until the probability of getting purple is $\frac{1}{2}$. How many purple counters does she add? *(3 marks)*

$P(purple) = \frac{1}{2}$

So $P(not\ purple) = \frac{1}{2}$

$P(not\ purple) = \frac{6+5}{x} = \frac{1}{2}$ ✓

$\frac{11}{x} = \frac{1}{2}$

$x = 11 \times 2 = 22$

There are many ways to model this. You could use a diagram, for example:

6	5	4	
6	5	4	?

So there are 11 purple counters altogether. ✓

11 – 4 = 7   Maya added 7 purple counters. ✓

7.2 Complementary Outcomes and Predicting Results

Many questions boil down to the outcome of an event being one result (A) or not being that result (not A). For example, when flipping a coin the outcome could be 'heads' or 'not heads' (tails). When you add the probabilities of all the complementary outcomes, the result is 1. It is certain that the result will either be A or not be A.
P(A) + P(not A) = 1

You can predict results using probability by multiplying the number of trials by the probability of the required outcome.

Example 7.2 📱

A set of cards have A, B, C or D written on them. A game is played by randomly picking a card, then replacing it in the pack.

The table shows the probability of picking a card with A, B or D on it.

> These questions might have a part a) 'find the missing probability' and a part b) 'use an existing probability to predict'. It is important that you can plan the steps when answering these questions yourself.

Card	A	B	C	D
Probability	0.43	0.2		0.09

Edward is going to model playing the game 75 times.

Approximately how many times should he expect to get a C? *(4 marks)*

P(not C) = 0.43 + 0.2 + 0.09 = 0.72 ✓
P(C) = 1 − P(not C) = 1 − 0.72 = 0.28 ✓
0.28 × 75 = 21 ✓

Edward would expect to get a C approximately 21 times. ✓

$$\begin{array}{r} 0.43 \\ 0.20 \\ +\,0.09 \\ \hline 0.72 \end{array}$$

$$\begin{array}{r} 1.00 \\ -\,0.72 \\ \hline 0.28 \end{array}$$

Marks are often lost on relatively simple calculations – written methods can help to reduce this risk. Take your time and show your working.

7.3 Using a Sample Space Diagram

A sample space diagram shows all the possible outcomes of an event or combination of events. Within each event, all possible outcomes must be equally likely. This is often used for combined events such as rolling a fair dice and spinning a fair spinner, etc.

Example 7.3

At a school fair, Roisin sets up a game in which a player spins the two spinners shown. They add together the two numbers obtained to get a final score.

Roisin will give a prize of £2.50 for a winning final score but she does not know what score she should use.

	2	3	5	6	7
1	3	4	6	7	8
3	5	6	8	9	10
3	5	6	8	9	10
4	6	7	9	10	11
5	7	8	10	11	12
7	9	10	12	13	14 ✓✓

Each edge of the spinner is equally likely, so you can set up a sample space diagram to show all the different (equally likely) outcomes. NB: there are two number 3s on the hexagonal spinner so they both need to be included.

a) What winning final score should Roisin use in order to maximise her profits? *(3 marks)*

Roisin could use any of the numbers that just appear once, as they are all equally as likely. 3, 4, 13 or 14. ✓

Any one of these answers would gain the final mark.

b) 120 people will play the game during the fair. If Roisin charges 20p a go, how much profit can she expect to make? *(3 marks)*

If this is on the non-calculator paper, take your time and show your written method of calculation where relevant.

Total income = 0.20 × 120 = £24 ✓

Number of prizes expected = 120 × $\frac{1}{30}$ = 4

Predicted prize money handed out = 2.50 × 4 = £10 ✓

Profit = 24 − 10 = £14.00 ✓

The probability of getting a winning score is $\frac{1}{30}$ as there are 5 × 6 possible equally likely results, of which only one is 'favourable'.

Show clearly what each calculation finds to make it easy for you to follow your own working. It will also help for checking and for the examiner to mark it.

7.4 Experimental Probabilities (and Predicting Outcomes)

When you flip a fair coin, each outcome – heads (H) or tails (T) – is equally likely. This is true every time you flip the coin, whatever the last result was. This means that the result HTHT, in that order, is just as likely as TTHT or HHHH.

If a sample or experiment does not have many results, it is unlikely to represent the theoretical probability (in this case 0.5). If you increase the number of results, then the experimental probability becomes a better representation of the theoretical probability.

Example 7.4

Jermaine is rolling a dice and gets the following results:

Score	1	2	3	4	5	6
Frequency	2	3	6	1	2	1

a) Jermaine says the dice is biased. Is he correct? Justify your answer. *(1 mark)*

Jermaine may be correct but he doesn't have enough evidence to support his conclusion at this time. There are relatively few results in this experiment. To be able to judge if the dice is biased, the experiment would need to be repeated many more times. ✓

Make sure you justify your answer.

b) Jermaine continues with the experiment. He rolls the dice until he has rolled it 400 times. He concludes (correctly) that his data supports that the dice is three times more likely to land on a 3 than on any other number, and that all other outcomes are equally likely.

How many times would you expect Jermaine to have rolled a 3? *(4 marks)*

Let x be the number of times you would expect a 1 to be rolled. So $3x$ is the number of times you would expect to get a 3. All other numbers (2, 4, 5, 6) are as likely as 1, so you would expect to get each of them x times.

Mark schemes will show different methods for answering questions. If you have a way that makes sense to you, use it. Show your working clearly so the examiner can understand it easily too.

Total number of rolls = 400

$400 = x + x + 3x + x + x + x$ ✓

$400 = 8x$ ✓

$x = 50$ ✓

$3 \times 50 = 150$

Jermaine is expected to have rolled a 3 approximately 150 times. ✓

7.5 Combined Events, Dependence and Tree Diagrams

When you consider more than one event, you need to consider the effect of one on the other. Sometimes the first event has no effect on the second but if there is dependence, the outcome of one event changes the probabilities of the second event. A **tree diagram** can be useful to clearly show what is happening.

Example 7.5 📖

A bakery sells trays of cupcakes with different flavoured fillings. Laura orders a tray with 7 lemon flavoured centres, 4 chocolate centres and 1 salted caramel centre. She takes a cupcake at random and eats it. She then takes a second cupcake.

a) What is the probability that Laura got two of her favourite lemon flavoured cupcakes? *(4 marks)*

The question doesn't specify what method to use. As there is a change in probability between the first and second event – a 'non-replacement' question – a **probability tree** is a good way to represent the situation.

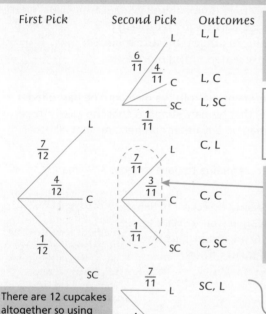

Using shorthand or initials can save time. Generally the question will have different first letters, making this easy. To be super clear, add a key.

Key:
L is lemon
C is chocolate
SC is salted caramel

You can sometimes save time by only filling in the relevant branches to the question. However, filling in all the branches can help you spot 'silly mistakes'. Check that the branches from each node sum to 1.

There are 12 cupcakes altogether so using fractions means you can avoid recurring decimals.

In this situation, you can simplify your tree slightly. If Laura got salted caramel on her first pick, there won't be salted caramel left on the second pick. Leaving it in place won't drop marks however, as long as it has a probability of 0.

$$P(L, L) = \frac{7}{12} \times \frac{6}{11} = \frac{42}{132} = \frac{7}{22} \checkmark$$

$\checkmark$

b) What is the probability that Laura got two cupcakes that were different flavours? *(4 marks)*

First Pick	Second Pick	Outcomes	Highlight the required branches on the tree diagram.

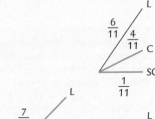
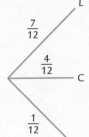

$P(L, C) = \frac{7}{12} \times \frac{4}{11} = \frac{28}{132}$

$P(L, SC) = \frac{7}{12} \times \frac{1}{11} = \frac{7}{132}$ ✓

$P(C, L) = \frac{4}{12} \times \frac{7}{11} = \frac{28}{132}$

$P(C, SC) = \frac{4}{12} \times \frac{1}{11} = \frac{4}{132}$ ✓

$P(SC, L) = \frac{1}{12} \times \frac{7}{11} = \frac{7}{132}$

$P(SC, C) = \frac{1}{12} \times \frac{4}{11} = \frac{4}{132}$ ✓

Don't simplify at this stage, even if you can, as your next step is to add the different combinations. In this case, as in most cases, the denominator will be the same along each branch, making the fraction sum easier.

$P(\textit{different flavours}) = \frac{28}{132} + \frac{7}{132} + \frac{28}{132} + \frac{4}{132} + \frac{7}{132} + \frac{4}{132}$

$= \frac{78}{132} = \frac{13}{22}$ ✓

7.6 Combining Probabilities and Algebra

In 'Show that' questions you have a final point to get to, but it isn't always obvious what the first step should be. Try not to be thrown by complicated looking algebra and start with what you know.

Example 7.6

There are 15 marbles in a jar. p marbles are blue and the rest are clear. Sophie takes two random marbles from the jar.

Show that the expression, in terms of p, for the probability that Sophie gets two marbles the same colour is $\frac{p^2 - 15p + 105}{105}$. *(7 marks)*

> If a question involves algebra and you are struggling to know how to approach it, try pretending that the algebraic term is a number. For example, if p was 2 could you answer this question? What would you do?

First Pick　　**Second Pick**

$$P(B, B) = \frac{p}{15} \times \frac{p-1}{14} \checkmark$$

$$P(C, C) = \frac{15-p}{15} \times \frac{14-p}{14} \checkmark$$

$$P(B, B) = \frac{p}{15} \times \frac{p-1}{14} = \frac{p(p-1)}{210}$$

$$= \frac{p^2 - p}{210} \checkmark$$

$$P(C, C) = \frac{15-p}{15} \times \frac{14-p}{14} = \frac{(15-p)(14-p)}{210} \checkmark$$

$$= \frac{210 - 29p + p^2}{210} \checkmark$$

$$P(\text{both the same}) = P(B, B) + P(C, C)$$

$$= \frac{p^2 - p}{210} + \frac{210 - 29p + p^2}{210} \checkmark$$

$$= \frac{2p^2 - 30p + 210}{210} = \frac{p^2 - 15p + 105}{105} \checkmark$$

> You can model this situation using a probability tree as shown.

> Even if both marbles are drawn simultaneously, model it as one then the other. In the first instance, the probability of getting a blue marble is found by saying there are p blue marbles out of 15 marbles. Having got one blue marble, there would be $p - 1$ blue marbles in the jar with a total of 14 marbles altogether. The same principle applies to the clear marbles.

7.7 Problem Solving, Algebra and Reasoning

Example 7.7

Sara is helping to hand out sweets at a charity event. There are N sweets in a bucket. 120 of them are strawberry flavoured. Sara takes 200 sweets randomly from the bucket. 23 of the sweets she has taken are strawberry.

a) Use this information to estimate the value of N. *(2 marks)*

Considering the sample, $\frac{23}{200}$ sweets are strawberry.

$\frac{23}{200} \approx \frac{120}{N}$ ✓

$N \approx \frac{120 \times 200}{23}$

$\approx 1043.47826...$

$N \approx 1043$ ✓

You can approximate the probability of getting a strawberry sweet in the sample to that of getting a strawberry sweet from the entire bucket.

Using specific mathematical notation is a quick and efficient way of communicating. In this case $\approx$ means approximately equal to.

It makes sense to round the answer to the nearest whole number.

Remember that approximations in mathematics are not just guesses. Calculations and assumptions should be made clearly.

b) There was exactly the same number of each type of sweet originally in the bucket.

i) Estimate how many different types of sweet there are. *(2 marks)*

$\frac{23}{200} = 0.115$ ✓

$1 \div 0.115 = 8.6956...$

There are approximately nine different types of sweet, based on the proportion of strawberry sweets found in Sara's sample. ✓

You are still assuming that the proportion of strawberry sweets in Sara's sample is roughly equivalent to that of the whole bucket. However, the additional information that there is an equal number of each type of sweet means that you will need to round.

ii) Use your answer to part **i)** to make a new estimate for the value of N. *(1 mark)*

$9 \times 120 = 1080$ ✓

You could estimate from part **i)** that there are approximately 1080 sweets in the bucket.

Probability

7.8 Approaching Questions from Different Angles

Examiners will always search for new ways of approaching questions to check you understand the subject material.

Example 7.8 🖩

A spinner is numbered 1 to 5.

John spins the spinner until he gets a 4.

a) Work out the probability that John spins the spinner:

   i)   exactly once. *(1 mark)*

$P(4 \text{ on first spin}) = \frac{1}{5}$ ✓	The probability that he spins the spinner once is the probability that he gets a 4 on his first spin.

   ii)   exactly twice. *(1 mark)*

$P(4 \text{ on second spin}) = P(\text{not } 4, 4)$ $= \frac{4}{5} \times \frac{1}{5} = \frac{4}{25}$ ✓	The probability that he spins the spinner exactly twice is the probability that he does not get a 4 on the first spin then does on the second.

   iii) more than twice. *(2 marks)*

$P(\text{more than 2 spins}) = 1 - P(\text{First}) - P(\text{Second})$ $= 1 - \frac{1}{5} - \frac{4}{25} = 1 - \frac{9}{25}$ ✓ $= \frac{16}{25}$ ✓	The probability that he spins the spinner more than twice is the probability that he did not just spin it once or twice.

You could use a diagram to model what is going on. This tree diagram shows the possible outcomes.

Alternative method

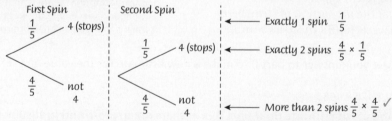

$P(\text{more than 2 spins}) = P(\text{not } 4, \text{not } 4)$

$= \frac{4}{5} \times \frac{4}{5} = \frac{16}{25}$ ✓

Having a second method can help you check answers. Also, if you are stuck, trying a different method might help.

b) Find an expression for the probability associated with spinning the spinner exactly n times. *(3 marks)*

1 time = $\frac{1}{5}$

2 times = $\frac{4}{5} \times \frac{1}{5} = \frac{4}{25}$

3 times = $\frac{4}{5} \times \frac{4}{5} \times \frac{1}{5} = \frac{16}{125}$ ✓

n times = $\left(\frac{4}{5}\right)^{(n-1)}$ ✓ $\times \frac{1}{5} = \frac{1}{5}\left(\frac{4}{5}\right)^{(n-1)}$ ✓ $= \frac{4^{n-1}}{5^n}$

The question is effectively asking for the nth term. Start by looking for patterns in the results you have so far.

Check the question carefully to see if the final answer needs to be in a certain form. It is generally good practice to simplify as far as possible. In this case it asks for 'an expression' so either of the final two answers would be acceptable, or any other equivalent expression.

7.9 'Simple' Venn Diagrams

A Venn diagram is good when there are options that are not **mutually exclusive**, i.e. things that can both be true at the same time. These are represented by the overlap sections in the diagram.

Example 7.9

A group of 32 students were asked if they like to watch football or rugby:

• 8 students didn't like to watch either.
• 19 students said they enjoyed watching football.
• 12 students said that they enjoyed watching rugby.

Use this information to draw a Venn diagram and find the probability that a student chosen at random likes to watch both football and rugby. *(3 marks)*

There are two options and a neither. You can set up your Venn diagram with two circles to represent football and rugby.

As 8 students don't like either, you know that there are 24 students who like either football, rugby, or both.

You can find how many like both by adding together those that like football and those that like rugby and taking away 24. This finds the number that must be in the overlap section.

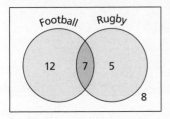

Having found the overlap, you can fill in all the remaining sections. Check all the relevant sections are right by considering how many are in each circle and that altogether there are 32.

✓✓   P(student likes both) = $\frac{7}{32}$ ✓

7.10 More Complex Venn Diagrams

Venn diagrams with more variables become a bit more complicated. However, you will be given more information to start with. Use what you know and treat it like the number puzzle it is. Don't be afraid of pencilling in things to help you think about it. Generally, the second part of a Venn diagram question may ask 'given that the person is x, what is the probability they are also y?'. In this case, make sure you adjust your denominator to represent the subgroup x rather than the entire group.

Example 7.10

At a wedding reception, the guests are served afternoon tea consisting of:
a scone, a pot of cream and a bowl of strawberries.
Of the 150 guests, everybody had something.
20 people didn't have cream.
67 people had all three items.
2 people had only cream.
109 people had strawberries.
113 scones were served.
25 people had strawberries and cream but no scone.

> Try not to get put off by the context too much. It would be easy to get distracted by why anyone would just have cream but in terms of the question it isn't important!

a) What is the probability that a guest chosen at random ate a scone with cream but no strawberries?
(4 marks)

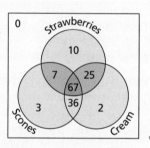

$P(\text{scone with cream}) = \frac{36}{150}$ ✓

> Out of all the guests (150), there are 36 who had a scone with cream but no strawberries.

✓ ✓ ✓

Step-by-step completion of the Venn diagram:

Straight from the question, fill in everything you know.	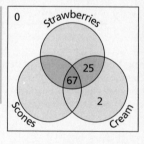	As 20 people didn't have cream, you know $150 - 20 = 130$ people did have cream. $130 - 67 - 25 - 2 = 36$

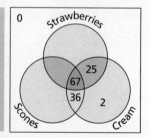

(continued on next page)

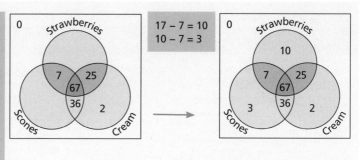

Finding the overlap: 109 people had strawberries, 109 − 67 − 25 = 17; 113 had scones, 113 − 67 − 36 = 10; 17 + 10 − 20 = 7 ∴ 7 had strawberries and scone.

17 − 7 = 10
10 − 7 = 3

b) Given that a guest had eaten a scone, what is the probability that the guest also had cream? *(2 marks)*

P(cream given scone) = $\frac{103}{113}$ ✓ ✓

If you know the guest had a scone, then the total number of guests being considered is now only 113. Of those 113 guests, 103 had cream.

7.11 Frequency Trees

A **frequency tree** shows the numbers involved in a combined event. If there are just two parts, then a two-way table can also be used.

Example 7.11

In a medical trial of a new headache cure involving 150 volunteers, x people were given a placebo and the rest were given the new drug. Of those given the placebo, 12 reported an improvement in the severity of their headaches. Of those who were given the new drug, 28 reported an improvement.

If there is a word in a question that you don't know, it might not matter. In this question it talks about a placebo; as long as you can see it is an option of that or the actual drug, you can still do the question.

a) Draw a frequency tree to represent these results. *(2 marks)*

Probability

You can draw frequency trees in different orders. The result should still be the same. Here are two possible layouts for this situation. Both have advantages over the other in terms of ease of doing and ease of interpretation.

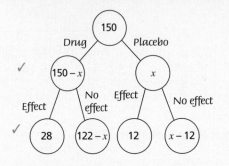

Alternative method

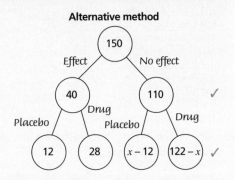

This question specifies a frequency tree, so a two-way table would earn no marks. Had it requested a two-way table, the answer is as shown here. If the question asks for a frequency diagram, or just a solution, then you can choose which diagram you use.

	Effect	No Effect	Total
Placebo	12	$x - 12$	x
Drug	28	$122 - x$	$150 - x$
Total	40	110	150

b) The probability of selecting someone who had received the placebo given that they felt no significant effect is 0.5

Find the value of x. *(2 marks)*

$\frac{x - 12}{110} = 0.5$ ✓

$x - 12 = 0.5 \times 110$

$x = 55 + 12$

$x = 67$ ✓

67 of the volunteers were given the placebo.

> If your diagram is either of the methods shown in part a), then the value 110 is easy to spot as the denominator.

Alternative method

$122 - x = x - 12$ ✓

$2x = 122 + 12$

$2x = 134$

$x = 67$ ✓

67 of the volunteers were given the placebo.

> There are always different ways of approaching questions. Alternatively, if the probability of having had the placebo given that they felt no effect is 0.5, that means the number of people who felt no effect with the placebo equals the number of people who felt no effect having had the drug.

 For more on the topics covered in this chapter, see pages 28–31 of the Collins Edexcel Maths Higher Revision Guide.

Probability: Key Notes

- Probability is used to measure, consider and communicate about chance and likelihood.
- The probability scale goes from 0 (impossible) to 1 (certain).

0   0.5   1

- If you find a negative probability or a probability greater than 1, you know there has been a mistake.
- Probabilities can be expressed as fractions, decimals or percentages. You should take your lead from the question wherever possible. <u>Do not</u> use ratios or 'out of'. If the question does not give you one, use whichever you are most comfortable with.
- There is a lot of specific terminology and notation – check the Glossary on page 139.
- Theoretical probability of any particular outcome to an event is
  $\frac{\text{The number of possible favourable outcomes}}{\text{The total number of outcomes}}$, where each outcome is equally likely.
  - A favourable outcome is the outcome that you are calculating the probability for.
- If you do an experiment or take a sample, it has to be large enough so the probabilities and proportions of the sample (or generated in the experiment) can be used to estimate the proportions and probabilities of the whole group (the theoretical probabilities).
- You can predict outcomes using probability by multiplying the probability of the outcome by the total number of times the event happens.
- Mutually exclusive outcomes are outcomes that cannot both happen at the same time, e.g. getting a vowel and getting the letter G when selecting one letter at random. You can add the probabilities of mutually exclusive outcomes to find the probability of getting either of them.
- The sum of the probabilities of mutually exclusive and exhaustive outcomes to a given event is 1 (it is certain that one of the outcomes will happen).
- The probability of a particular outcome happening is equal to (1 – the probability of that outcome not happening). P(A) = 1 – P(not A)
- There are lots of diagrams you can use. Be careful to use the right one if it is stated in the question (especially note the difference between a frequency tree and a probability tree):
  - A sample space shows all the possible outcomes of an event. It can be used for a combination of two events but tends to get complicated beyond that.
  - Lists, tables, grids and trees can be used to represent combined events. They can be good for **conditional probabilities**.
  - Frequency trees show the frequencies of a particular outcome at the ends of the branches. You read the frequencies and use them to find the probabilities.
  - Probability trees show the outcome at the end of each branch and the probability on the branch. They are very useful for events that are dependent, but can still be used if events are **independent**.

8 Statistics

Statistics is used widely to help understand situations. It involves collecting, analysing and representing data. Almost every job will involve some sort of statistics. For example, sports coaches look at performance data, doctors need to know about the effectiveness of treatments and businesses write reports for shareholders. It is often said that it is easy to lie with statistics and therefore it is very important to have a good understanding to avoid being misled.

8.1 Populations and Sampling

The population in statistics refers to everyone or everything that you are interested in. For example, that could be all the students at your school, all men aged 20 to 25 in the UK or it could refer to things such as all the books in a library. If you want to find out information about a population, it is often unrealistic to use the whole population so a sample will be used.

Sampling Methods

A classic question is to be given a sampling method and asked whether it is a good one. It usually isn't, so you will need to explain the potential problems.

Example 8.1.1

Amy wants to find out about how many CDs people buy, so she stands outside CD World on a Tuesday in December at 11 am and asks every tenth person how many CDs they have bought this week. Evaluate this sampling method. *(2 marks)*

Amy's location means that many of the people she is asking will be customers of CD World and so likely to buy CDs regularly.

At 11 am on a weekday, many people will be at work or school, so her sample will not be representative.

In December, people could be buying Christmas presents, so their shopping habits may be different from a normal week.

By asking every tenth customer, Amy is taking a systematic sample, which means that the population should be evenly sampled. ✓ ✓

You are looking for any sources of bias. Try to think about parts of the population that may be missed out or over-represented.

There are two marks available, so you will need to make two points. They must be sufficiently different, so you cannot say some people will be at work and some people will be at school as these points are both highlighting a problem with the time of the study.

Capture/Recapture

This type of question is new to GCSE maths but the method is widely used by scientists. It uses proportion to estimate the size of a population based on a sample.

Example 8.1.2

A gardener wants to find out how many snails live on her allotment. She catches 50 snails one day and marks them with a pen, before releasing them. The next day, she catches 30 snails and 12 have a mark on them.

Work out an estimate for the total number of snails on the allotment. State any assumptions you have made. *(4 marks)*

$$\frac{\text{Total number of snails }(T)}{\text{Number captured the 1st time}} = \frac{\text{Number captured the 2nd time}}{\text{Number of tagged snails captured 2nd time}}$$

The proportion of marked snails in the second sample should be the same as the proportion of the first sample to the total number of snails.

$$\frac{T}{50} = \frac{30}{12} \checkmark$$

$$T = 50 \times \frac{30}{12} \checkmark$$

$$T = 125 \checkmark$$

Showing your method is important here. You might have a different method and that is fine – just make it clear what you are doing and why.

Assumptions could include:

- *The mark does not wear off.*
- *The population remains the same.*
- *The sample is random each time.* ✓

You have not finished yet! The last bit of the question asks about assumptions made; students often forget this bit and lose an easy mark. Any of the assumptions given would get you the mark. The assumptions will be very similar for every question but should refer to the context given.

8.2 Measures of Location and Spread

There are times when it is useful to have summary statistics which give a quick, concise representation of the data. The three averages (**mean, median** and **mode**) and the **range** give information about the location of the data and how spread out it is. Often these measures are combined as part of a question about displaying data, but they can also be made into a higher level question by requiring you to work backwards from information given. If the data has been grouped, then additional steps are needed in order to find the averages from the table given.

Statistics

Quartiles and the **interquartile range** are also used to measure location and spread, but they are generally used in the context of either **cumulative frequency** curves or **box plots**.

Example 8.2.1 (finding unknowns)

There are eight children who have an average age of 14 years old. A pair of twins, Adam and Betty, join the group and the average age decreases to 13 years old. How old are Adam and Betty? *(3 marks)*

The total age of the eight children is 8 × 14 = 112

When the twins join the new total is 10 × 13 = 130 ✓

The average takes the total and divides it by the number of values, so you do the opposite here to find the total by multiplying the average by the number of children.

They are twins so must be the same age, which can be called x.

$2x + 112 = 130$ ✓

$2x = 18$

$x = 9$ so Adam and Betty are 9 years old ✓

You can set up an equation to solve, with x as their age.

Make sure the answer is obvious at the end.

The algebra makes it clear but you could choose to do this without; just make sure you write down the steps you use.

Example 8.2.2 (averages from a table)

Mr Nyembo asks his form group how far they live from the school. The information is shown in the table.

Distance (d miles)	Frequency (f)
$0 < d \leqslant 1$	12
$1 < d \leqslant 2$	8
$2 < d \leqslant 5$	7
$5 < d \leqslant 10$	5

There are a number of different averages questions you can be asked from a table. You wouldn't usually get so many sections but this example shows you some of the types of questions that could be asked.

The data has been grouped into categories, which makes it easier to put into a table. But it is no longer possible to know the exact distances the children live from the school, so the mode and the median will be given as a class and the mean will be an estimate.

a) Write down the modal **class interval**. *(1 mark)*

The modal class is $0 < d \leqslant 1$ miles. ✓ | The **modal class** is the category with the highest frequency. Make sure you state the category, not the frequency, as that is a common mistake.

b) Find the median class interval. *(2 marks)*

Distance (d miles)	Frequency (f)	C.F.
$0 < d \leqslant 1$	12	12
$1 < d \leqslant 2$	8	20
$2 < d \leqslant 5$	7	27
$5 < d \leqslant 10$	5	32

This category goes from 13 to 20 so the 16.5th term will be here.

To find the median, start by adding an extra column for the cumulative frequency if it hasn't already been given. You can do this separately underneath but it makes sense to keep things together and the examiner will be used to seeing this method.

You find which term will be the median using the formula $\frac{n+1}{2}$. The cumulative frequency column of the table shows the last value in each category and is used to find which category has the median.

$\frac{n+1}{2} = \frac{32+1}{2} = 16.5\text{th term}$ ✓
The median class is $1 < d \leqslant 2$ miles. ✓

The estimated mean distance students live from the school is 2.5 miles. A student was absent when Mr Nyembo collected his data. This student lives 5.5 miles away from the school. Anka and Brad work out the new estimated mean.

Anka	Brad			
$\frac{2.5 + 5.5}{2}$	Distance (d miles)	Frequency (f)	Midpoint (x)	fx
	$0 < d \leqslant 1$	12	0.5	$12 \times 0.5 = 6$
$= \frac{8}{2}$	$1 < d \leqslant 2$	8	1.5	$8 \times 1.5 = 12$
	$2 < d \leqslant 5$	7	3.5	$7 \times 3.5 = 24.5$
$= 4$	$5 < d \leqslant 10$	5̶6	7.5	$6 \times 7.5 = 45$
	Totals	33		87.5
	$\frac{87.5}{33} = 2.65$ to 2 d.p.			

c) Whose working out is correct? You must explain your answer. *(2 marks)*

Brad is correct because he has put the value into the correct category before working out his estimate of the mean. ✓✓

When working out the mean, check that your answer makes sense in the context of the question. Instead of dividing by the total frequency, some students divide by the number of categories (which will give a very large answer, so you will know you have gone wrong).

8.3 Displaying Data

Scatter Graphs

Scatter graphs are used to look for relationships between two sets of data. The data will be in pairs so can be plotted like coordinates and the graph is used to look for a **correlation** between the two sets of data.

Example 8.3.1

Anna keeps chickens and records how old each chicken is and how many eggs they each lay in a week. The information has been recorded in a table and is shown on the scatter graph.

Age of Chicken (months)	Number of Eggs per Week
7	6
18	5
20	4
35	2
9	6
11	5
17	4
14	6
28	3

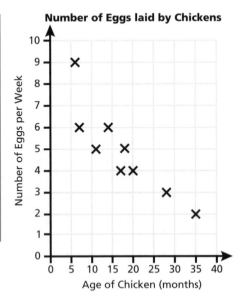

Number of Eggs laid by Chickens

a) Anna made a mistake when plotting one of the points. Correct this mistake. *(1 mark)*

Shown on the diagram. ✓

b) Describe the relationship between the age of chickens and the number of eggs per week. *(1 mark)*

As the age of the chicken increases, the number of eggs laid per week decreases. ✓

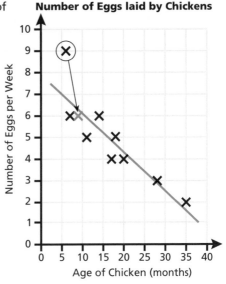

Number of Eggs laid by Chickens

It is important to read the question carefully here as you are being asked for the relationship, not the correlation. When describing relationships, you need to use the context in your description. It is fine to also say that it is a negative correlation, but it is the context that will get you the mark.

c) Estimate the number of eggs that would be laid by a chicken that is 38 months old and comment on the reliability of your prediction. *(2 marks)*

From the graph it would be estimated that 1 egg would be laid by a 38 month old chicken. ✓ Method

This prediction is extrapolated beyond the range of given values and so may not be very reliable. ✓ Comment

When making predictions, it is a good idea to use a line of best fit. Use your pencil and ruler to draw the line of best fit, then go up to the line from the value you are given and across to find the prediction. Strong students will write a full explanation using the correct terminology.

Pie Charts

Pie charts show proportions. Each section represents the proportion of the whole. You will need your protractor for pie chart questions to ensure you measure the angles correctly.

Example 8.3.2

Harrison collected data about the types of fruit students chose at a revision day. He has started to put together a table and pie chart.

Complete the table and pie chart to show all of the information. *(5 marks)*

Statistics

Type of Fruit	Number of Students
Apple	30
Banana	
Grapes	25
Orange	15

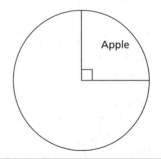

Initially this question looks like there isn't enough information but if you start off as you would for a normal pie chart question and fill in everything you know, then you can start to work out the rest.

$\frac{1}{4}$ of the students chose an apple

30 × 4 = 120 students in total ✓

30 + 25 + 15 = 70

120 − 70 = 50 chose a banana ✓

Type of Fruit	Number of Students	Angle
Apple	30	90°
Banana	50	$\frac{50}{120} \times 360 = 150°$
Grapes	25	$\frac{25}{120} \times 360 = 75°$
Orange	15	$\frac{15}{120} \times 360 = 45°$
Total	120	360°

✓ Method shown

It is helpful to add an extra column to the table given to put in the angles and to add a row at the bottom for the totals. From the pie chart you can see that apple is 90°, so you know that this represents 30 students. As 90° is a quarter of 360°, you can work out the total number of students and so find how many chose a banana. When you have worked out the angles, do a quick check that they add up to 360°.

Orange

Apple

45°

Grapes
75°

150°

Banana

✓ Angles correct
✓ Drawn accurately

Draw the sections using a pencil, ruler and protractor. Don't assume that your last section is correct; measure it to check so that if you have made a mistake earlier on you can fix it. Your answers will be scanned in so there is no point using colours as the examiner will not see them and it can make things very difficult to read. Instead, label each section or use simple patterns to represent each section and show this on a key.

Box Plots

A box plot is a quick way to show information about a data set and is sometimes called a box and whisker diagram. The diagram below shows what each line represents.

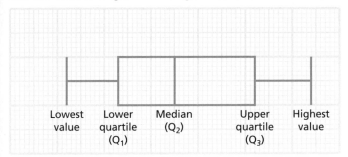

| Lowest value | Lower quartile (Q_1) | Median (Q_2) | Upper quartile (Q_3) | Highest value |

Example 8.3.3

Here are the weights of 15 female barn owls in grams:

372, 343, 397, 365, 323, 334, 370, 358, 349, 341, 381, 361, 374, 360, 352

a) Draw a box plot to illustrate the information given. *(3 marks)*

323, 334, 341, 343, 349, 352, 358, 360, 361,
365, 370, 372, 374, 381, 397

$Q_1 = \frac{n+1}{4} = \frac{15+1}{4} = $ 4th term = 343

$Q_2 = \frac{n+1}{2} = \frac{15+1}{2} = $ 8th term = 360

$Q_3 = 3\left(\frac{n+1}{4}\right) = 3(4) = $ 12th term = 372 ✓

The numbers have not been given in order so the first thing you need to do is rewrite them in ascending order.

You can work out the median and quartiles by counting along from each edge, but it helps to write out your method so that your work can be easily checked and your method is clear.

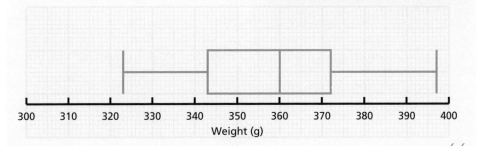

✓✓

Statistics

b) The box plot below shows the distribution of the weights of male barn owls.

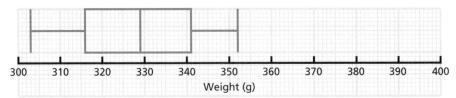

Weight (g)

Use the box plot to compare the distributions of the weights of the female barn owls with these male barn owls. *(2 marks)*

The median for the female owls is 360 g, which is higher than the males' median of 329 g. This shows that, on average, the female owls are heavier than the male owls. ✓

The interquartile range of the male owls is smaller at 25 g compared to an interquartile range of 29 g for the females.

The range of the male owls is smaller at 49 g, compared to a range of 74 g for the females. ✓

When making comparisons, you need to be specific and say which one is larger or smaller. Use figures to back up any claims you make and refer to the context of the question. There is one mark for comparing the medians and the other can be for comparing either the interquartile range or the range.

Histograms

Histograms are used for continuous data where the class widths are not all equal. It is the area of the bar, rather than the height, which shows how many items are represented. You will need to be confident both drawing histograms and being able to work backwards to find frequencies. You also need to be able to use histograms to find estimates.

Example 8.3.4

Adeel is researching song lengths so takes a sample of 290 pop songs and finds the length in seconds.

Length of Song (s)	140–180	180–200	200–210	210–220	220–240	240–280
Frequency			50	40	20	20

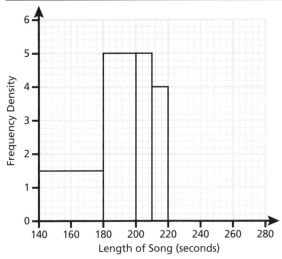

a) Complete the histogram. *(2 marks)*

For 220–240 $\frac{20}{20} = 1$ ✓

For 240–280 $\frac{20}{40} = 0.5$ ✓

> Use the formula **Frequency density** $= \frac{\text{Frequency}}{\text{Class width}}$ to work out the frequency density for the two bars that need to be drawn. Make sure you use a pencil and ruler – accuracy is important and if you make a mistake, you can make changes.

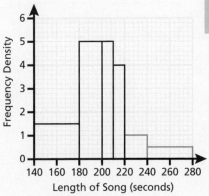

b) Complete the table. *(2 marks)*

Frequency = Frequency density × Class width

For 140–180 $1.5 \times 40 = 60$ ✓

For 180–200 $5 \times 20 = 100$ ✓

> To find the frequency you are using the same formula, just rearranged. From the graph you can read off the frequency densities.

c) Work out an estimate for the number of songs longer than 3 minutes 45 seconds. *(3 marks)*

> Make sure you check the context and the units. The table and histogram use seconds and this question asks about minutes, so first you need to convert to seconds. 225 seconds fall part way through the 220–240 category. It is 5 seconds into a 20-second category, so you need the remaining $\frac{3}{4}$ of the category.

    3 *minutes* 45 *seconds* is 3 × 60 + 45 = 225 *seconds* ✓

    $\frac{3}{4}$ × 20 = 15 ✓

    15 + 20 = 35 ✓

Time Series Graphs

A time series graph looks at patterns over time and is particularly useful for data that might be seasonal, where you are looking for trends. For example, ice-cream sales are likely to be much higher in summer months so comparing sales in December to those in June wouldn't be very helpful, but looking at sales for three years split up by season could help to identify trends.

Example 8.3.5

Emma works in an office and wants to find out about how many people walk or cycle to work. She has collected information over two years and wants to present it as a time series graph.

	Spring (2015)	Summer (2015)	Autumn (2015)	Winter (2015)	Spring (2016)	Summer (2016)	Autumn (2016)	Winter (2016)
Walk	40	41	39	33	39	44	41	36

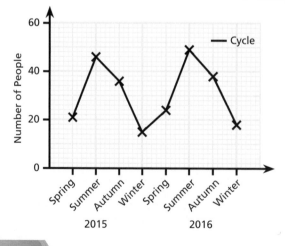

a) Draw on the information for people who walk to work. *(3 marks)*

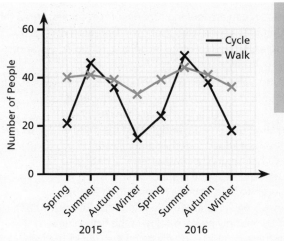

Plot the points in line with the labels for the seasons. Use a pencil and have a rubber ready in case you go wrong. The points should then be joined with straight lines using a ruler. Don't forget to include a key.

✓ Points plotted

✓ Points joined by lines

✓ Key

b) At which point is the difference between the number of cyclists and the number of people walking the highest? *(2 marks)*

Spring 2015	40 − 21 = 19
Winter 2015	33 − 15 = 18
Spring 2016	39 − 24 = 15
Winter 2016	36 − 18 = 18

The biggest difference is in spring 2015 when there were 19 more people walking than cycling. ✓ ✓

From the graph you can see that the biggest differences are in spring and winter so these are the differences to check. You might not need to write out each of the calculations if it is clear on the graph, but it is often helpful and makes sure your working out is clear.

c) Describe the trends in the number of people cycling and walking to work. *(2 marks)*

The number of people walking to work remains similar over the year with a small drop in the winter. ✓

The number of people cycling in 2016 is slightly higher for each season than in 2015. The number of people cycling to work is much lower in the winter than in the summer. ✓

There are two marks for this question, so you should be making at least two comparisons. It is a good idea to make a comparison for each data set if two data sets are given, so here there would be one mark for writing about the number of people who walk and another for those who cycle.

 For more on the topics covered in this chapter, see pages 80–83 of the Collins Edexcel Maths Higher Revision Guide.

Statistics: Key Notes

- If asked about a sampling technique, you are looking for any sources of bias. Try to think about parts of the population that may be missed out or over-represented.
- To estimate population size using a capture/recapture technique, set up an equation:

$$\frac{\text{Total number in the population } (T)}{\text{Number captured the 1st time}} = \frac{\text{Number captured the 2nd time}}{\text{Number captured the 2nd time that are tagged}}$$

- Data can be discrete (taking distinct values which could be numbers or categorical, e.g. red, blue, orange, etc.) or continuous (taking any value over a continuous range).
- The range measures how spread out data is (Largest value – Smallest value).
- The median is found by putting all values in order and finding the middle ($\frac{n+1}{2}$th) term. If there are two middle terms, they are added together and halved.
- The mode is the value (or values) that occurs the most often in a data set.
- The mean is the total of all values divided by the number of values.
- Add a cumulative frequency column to help find the median from a table.
- If the data is grouped, the mean will be an estimate (use midpoints for the x value).
- The mean from a table is found by multiplying each value in the first column by its frequency. Then find the sum of these products and divide by the sum of the frequencies. This can be written as Mean = $\frac{\Sigma fx}{\Sigma f}$
- To find the **lower quartile**, look for the $\frac{n+1}{4}$th term and for the **upper quartile** look for the $3(\frac{n+1}{4})$th term.
- The interquartile range is the difference between the upper and lower quartiles.
- There are a number of ways data can be displayed. Always use a pencil and ruler to draw graphs. Make sure you have labels on the axes and a key if necessary.
- Bar charts are used for discrete data; the height of the bars shows the frequency and the bars do not touch.
- Scatter graphs show the relationship between two variables. If you are asked to describe the relationship, your answer should be in the context of the question, e.g. as the temperature increases so do ice-cream sales. If it asks for the type of correlation, you should say whether it is positive (both variables are increasing), negative (as one increases, the other decreases) or if there is no correlation.
- When estimating from a scatter graph, **interpolation** is within the range of given values so is likely to be reliable. **Extrapolation** is outside the range so may be unreliable.
- Pie charts show the proportion of each category. Each section is a proportion of 360°.
- In frequency polygons, the frequency for each group is plotted at the group's midpoint.
- Cumulative frequency graphs show the running total. Points are plotted at the last value in the group. Your final point should be the same as the total number of items.
- Histograms are used for continuous data. The area of each bar represents the frequency and the y-axis shows the frequency density. You need to know Frequency density = $\frac{\text{Frequency}}{\text{Class width}}$
- Box plots show the summary statistics for a set of data. The first vertical line is the smallest value; the box shows the lower quartile, median and upper quartile; and the final vertical line is the largest value.
- Time series graphs show trends over time. The points are plotted in line with each label and are joined with straight lines.

9 Transferable Skills

Mathematics is valued not just for the techniques and procedures learned but also for the transferable skills that are developed. This section looks at some general exam hints and tips, as well as the types of questions in which you will need to use the mathematics you have learned in different contexts. These are the AO3 questions that will require you to use skills from different areas of mathematics. Most importantly, you will need strong reasoning skills and will need to give clear explanations.

9.1 Exam Hints and Tips

- Read the question before, during and after answering.
- Stay calm. You won't be asked to do things that are impossible (though hype on social media may lead you to question this!). Remember, there are a lot of marks on every question for method and intermediate answers, so do what you can, even if you don't know how it might help you reach the final answer.
- Always consider earlier parts to questions and whether they make later parts simpler.
- If the question needs multiple steps to solve it, write a brief plan of what you intend to do. This can help keep you on track or think about what you are trying to achieve in smaller steps, rather than considering the entire thing in one go.
- If the question has a diagram, add all the information to it as you go along. Redraw diagrams if they are getting over-complicated to show the part you need to focus on.
- Mark your answer in some way (such as a pencilled star on the page) if you feel it is worth coming back to, if there is time. Fight the urge to cross something out if you think you have made a mistake.
- You should be spending roughly one minute per mark; some will take less time and some more. Don't spend too long trying to get one or two early marks and then run out of time for big questions later on.
- When you think you have finished the question, do a final check:
  - Make sure you have answered the question. It is easy to find x, write it on the dotted line and move on, but were you asked to find x or was that an intermediate step?
  - Does the question ask you to round to a specific number of decimal places or significant figures?
  - Is your answer in the correct form?
  - If the question says to give your answer in its simplest form, there will be an easy mark for this so make sure you have simplified fully. This is especially important for fractions, algebra and surds.
  - Check the context of the question to see if your answer makes sense. If it doesn't, you probably need to check your working out.

9.2 Answering AO3 Questions

AO3 questions can be tricky to answer, often because they ask you to use more than one area of mathematics and your common sense (as the questions will generally be set in a context). There will usually be more than one way to answer the question, but in all cases having clear, easy-to-follow methods will be very important.

Translating Problems in Contexts into Mathematical Processes

Some questions require you to translate problems in mathematical or non-mathematical contexts into a process or series of mathematical processes. You will be given a situation from which you need to draw out the mathematics. You may be given more information than you need and you will have to work out which bits are important.

Example 9.2.1

Callum has an appointment at Binary Road, which is 8 miles from his house. His appointment is at 11am. He takes 10 minutes to walk to the Latitude Lane bus stop, which is half a mile from his house. When he gets there at 10:10, he sees the sign below.

Broad Street	09:14	09:36	09:58	10:19
Latitude Lane	09:29	09:51	10:13	10:34
Chapel Street	09:37	09:59	10:21	10:42
White Road	09:50	10:12	10:34	10:55
Binary Road	10:04	10:26	10:48	11:09
Garden Row	10:12	10:34	10:56	11:17
South Crescent	10:28	10:50	11:12	11:33

Buses are diverted via Division Street so will take up to an extra 15 minutes.

Apologies for the delay.

Callum goes home to get his bike so that he can cycle to Binary Road at 12mph. Should he do this instead? *(2 marks)*

Bus: 10:48 + 15 minutes = 11:03
Walking home: 10 minutes

Cycling:
$T = \frac{D}{S} = \frac{8}{12} = \frac{2}{3}$ hours

$\frac{2}{3} \times 60 = 40$ minutes ✓

Total 50 minutes so he would arrive at 11am if he cycled.

He should cycle because he will get there at 11am, just in time for his appointment. ✓

Label each part of the journey to avoid any confusion.

The information about the bus can be read from the table. Use your ruler to help you read from the table.

Don't forget to include the time taken for him to walk back home to get his bike.

Your final line should answer the question given, with reference to the calculations you have done.

Making and Using Connections between Different Parts of Mathematics

Some questions require you to make connections between different parts of mathematics and then use these connections. Links are made between different areas of mathematics to make you think about things in a different context. Algebra comes into almost all areas of mathematics and Chapter 4 contains many more examples of connections. Usually the individual steps will be fairly straightforward, but knowing where to start can be difficult. A good idea is to look at everything you know about the situation and have a go at something, but be ready to try a different tactic if your first doesn't work.

Example 9.2.2

The circle shown below has an area of 12π units^2. The centre of the circle is at the origin and the points A, B, C and D represent the intersections of the circle with the coordinate axes.

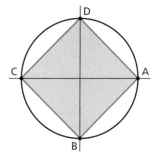

Find the perimeter of the quadrilateral ABCD. Give your answer in the form $a\sqrt{b}$, where a and b are integers to be found. *(4 marks)*

Draw information on your diagram as you find it to help you keep track.	$\dfrac{\pi r^2}{\pi} = \dfrac{12\pi}{\pi}$ $r^2 = 12$ $r = \sqrt{12}$ ✓	This question combines algebra, geometry and number skills.

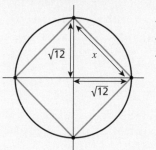

$x^2 = (\sqrt{12})^2 + (\sqrt{12})^2$

$x^2 = 12 + 12$

$x = \sqrt{24}$ ✓

Remember the perimeter is all four sides. Because the circle has its centre at the point (0, 0), the quadrilateral must be a square (so all four sides will be the same length).

$P = 4 \times \sqrt{24}$ ←

$= 4 \times \sqrt{4} \times \sqrt{6}$ ✓

$= 4 \times 2 \times \sqrt{6}$

$= 8\sqrt{6}$ ✓ ←

Make sure your answer is in the correct form: the surd should be fully simplified.

Transferable Skills

Evaluating Methods Used and Results Obtained

When you are evaluating, you are considering the good and bad points of the methods and results.

Example 9.2.3

Alena is saving money. She decides that each week she will check the serial number of the first £5 note she gets: if it ends with a prime number, she will spend it; otherwise she will put it into her savings jar.

She says that by the end of week 50, she will have saved £150. Her calculations are shown here. Explain whether she is correct. *(2 marks)*

$$0\ 1\ 2\ \cancel{3}\ 4\ \cancel{5}\ 6\ \cancel{7}\ 8\ 9$$
$$P(\text{not prime}) = \frac{6}{10} = \frac{3}{5}$$
$$\frac{3}{5} \times 50 = 30$$
$$30 \times £5 = £150$$

Alena's calculations are correct; however this is an estimated value. She is unlikely to get exactly 30 notes with a non-prime final digit, so she will probably have more or less than this amount. ✓ ✓

A good answer will be written in full, coherent sentences to get the point across well.

Interpreting Results in the Context of the Given Problem

Context is important in a lot of questions. You should refer to the context in order to check that the answer makes sense. In these cases, you will need to use your results to solve a problem or answer a question.

Example 9.2.4

Paul is a landscape gardener and he has planned a new, regular hexagonal raised bed as the centrepiece of a park. Each side will be 153 cm long and the distance between the parallel sides will be 265 cm. The raised bed will be 70 cm high and filled up to 10 cm from the top with a mixture of compost and topsoil in a ratio of 5 : 3. Compost costs £67.50 per cubic metre bag, or £35.50 for half a bag, plus VAT at 20%. The topsoil is free.

Paul has a budget of £200. Will this be enough? *(6 marks)*

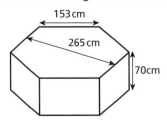

Redrawing the hexagon means that you can put on all the details that you know and decide what other information is needed. There are a few ways you could find the area; here, it has been split into two equal trapezia.

$$\text{Angle } \alpha = \frac{(6-2)180}{6} = 120°$$

$$\text{Angle } \theta = 120° - 90° = 30°$$

$$y = \tan(30) \times 132.5$$

$$= 76.49... \checkmark$$

$$x = 153 + 2(76.49...)$$

$$= 305.997... \checkmark$$

$$\text{Area} = 2\left(\frac{153 + 305.997...}{2}\right) \times 132.5$$

$$= 60\,817.21...$$

$$\text{Volume} = 60\,817.21... \times 60$$

$$= 3\,649\,032.68... \checkmark$$

The raised bed is a prism so, in order to find the volume, you need to know the area of the hexagon.

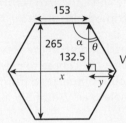

153
265   α θ
132.5
x y

In m^3 : $V = 3\,649\,032.68... \div 100^3$
$= 3.65\,m^3$ to 2 d.p. $\checkmark$

Be careful when converting units. It would be easy to just divide by 100, but it is a volume so it will be 100^3.

Compost : Topsoil   Total
$\times\frac{73}{160}$ 5 : 3 $\times\frac{73}{160}$ 8 $\times\frac{73}{160}$
2.28 : 1.37   3.65

The multiplier for the ratio isn't immediately obvious here, so can be found by dividing the total needed by the total of the given ratio.

2 bags at 67.50: 2 × 67.50   = £135.00
1 bag at 35.50: 1 × 35.50   = £35.50
£170.50 $\checkmark$

VAT: 170.50 × 1.2 = £204.60

The answer you get will be close to the amount given. If it isn't, go back and check your working out.

No, he doesn't have enough. $\checkmark$

Write out your final statement to answer the question: say whether Paul has enough money or not.

There are a lot of steps to this question so, when you finish, check to make sure that your working out is clear.

Evaluating Solutions and Identifying Assumptions

Some questions require you to evaluate solutions to identify how they may have been affected by assumptions made. Assumptions are things that have been taken to be true, but which may not necessarily be true or could have been interpreted differently. You are trying to find and explain any potential problems.

Example 9.2.5

Ala lives in a shared house with three other people and is in charge of the gas bill. They share the cost of the gas equally. There are two tariff options available, shown below.

Ala has the meter readings from the last day of each month in the previous year. They use more gas in the winter than in the summer but choose to pay a fixed amount per month, which is based upon the average amount of fuel used per month so that the extra cost in the winter is spread over the whole year. VAT of 5% is charged on all energy bills.

Initial Reading	1781
January	2591
February	3257
March	3815
April	4283
May	4769
June	5147
July	5453
August	5741
September	6083
October	6515
November	7019
December	7595

Deal 1

11.55p standing charge per day
9.55p per unit used

Deal 2

12.15p standing charge per day
8.75p per unit used

How much would each person in the house save per month by being on the cheaper deal? State any assumptions you have made. *(5 marks)*

Amount used over the year: 7595 − 1781 = 5814

If there are two or more things to compare, it can help to split the page so that calculations are clearly separated.

Deal 1

Cost for units used
5814 × 9.55 = 55 523.7

Standing charge
365 × 11.55 = 4215.75

Total
55 523.7 + 4215.75 = 59 739.45

VAT
59 739.45 × 1.05 = 62 726.4225p
= £627.26
(per year)

627.26 ÷ 12 = £52.27
(per month)

52.27 ÷ 4 = £13.07
(per person)

Deal 2

5814 × 8.75 = 50 872.5

365 × 12.15 = 4434.75

50 872.5 + 4434.75 = 55 307.25 ✓

55 307.25 × 1.05 = 58 072.6125p
= £580.73

580.73 ÷ 12 = £48.39 ✓

48.39 ÷ 4 = £12.10 ✓

Check the units. In this case it has been in pence so far, but now it makes sense to change to pounds.

Your two answers should be similar. It would be very unusual to get two very different answers so, if you do, check your working out carefully.

Difference 13.07 − 12.10 = 97p

They can save 97p a month each by being on Deal 2. ✓

It is assumed that they will use the same amount of gas next year.

It is assumed that prices will remain fixed for the year.

You may have worked this out differently, in which case you would have different assumptions.

It is assumed that there are 365 days, so it is not a leap year. ✓

Revision Sheet and Formulae to Learn

A formulae page will not be given in the exam paper, so any formulae will be given as part of the question. The notes below will help you avoid some common pitfalls.

Formulae You Will Be Given

Curved surface area of a cone = $\pi r L$

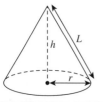

Cones

When using these formulae remember that h is the perpendicular height of the cone and L is the length of the slant height. It might be that you need to use trigonometry or Pythagoras' theorem to find L or h if you are not given it to start with.

Volume of a cone = $\frac{1}{3}\pi r^2 h$

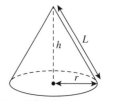

If you are finding the surface area of a solid cone, remember the base. This formula only finds the curved surface.

If you are already given the area of the circle, you can use it directly in this formula without having to calculate r.

Surface area of a sphere = $4\pi r^2$

Spheres

You may need to find the surface area or volume of a partial sphere, in which case it will be a proportion of the full sphere.

Volume of a sphere = $\frac{4}{3}\pi r^3$

Equations of linear acceleration/motion (SUVAT equations)

These equations are used when an object is travelling in a straight line. Think about the direction! If something is a negative value, it is in the opposite direction to a positive. If a car is moving at 5 m/s and accelerating at −1 m/s², it is slowing down (decelerating).

$v = u + at$	$s = ut + \frac{1}{2}at^2$	$v^2 = u^2 + 2as$	Where a is constant acceleration, u is initial velocity, v is final velocity, s is displacement from the position when $t = 0$ and t is time taken.

Formulae You Need to Know

These are some of the key formulae that you need to know. There are some useful suggestions for remembering formulae in Chapter 2 but plenty of practice is needed.

Area and Volume

Area of a parallelogram	Area of a trapezium	Area and circumference of circles
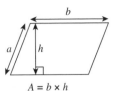 $A = b \times h$	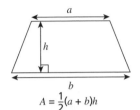 $A = \frac{1}{2}(a + b)h$	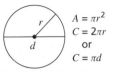 $A = \pi r^2$ $C = 2\pi r$ or $C = \pi d$
When using this formula, make sure you have the height and base. Often the length of the sloped side will be given as well to confuse you.	Trapezia can be drawn in any orientation, but the height will always be the distance between the two parallel lines.	Sometimes students are tempted to use 3.14 for π but you should always use the π button on your calculator. On a non-calculator paper, you will usually be asked to leave your answer in terms of π so you should simplify as much as possible.

Volume of a prism	Volume of a cylinder	Volume of a pyramid
cross-section $V = $ Cross-section $\times l$	$V = \pi r^2 h$	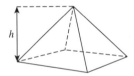 $V = \frac{1}{3} \times$ Area of base $\times h$
Don't try to do everything in one go here. Find the area of the cross-section first and then multiply by the length. A cylinder is a special type of prism so the same principle is used.		Check that you have the height and not the length of a side. You might have to use Pythagoras' theorem to first find the height.

Trigonometry

For all trigonometry questions, you should check your calculator is in the correct mode. It is good practice to reset your calculator at the start of the exam and you should know how to do this without instructions in case you accidentally put it into a different mode in the exam.

Anytime you put in a trigonometric function of an angle (sine, cosine or tangent), your calculator will automatically put in a bracket. Remember to close this bracket once you have keyed in the angle. If you get an error, it is usually because you are trying to do sine^{-1} or cosine^{-1} of a number greater than 0. Go back and check your working out.

Revision Sheet and Formulae to Learn

Pythagoras' theorem

$$a^2 + b^2 = c^2$$

Double check whether it is the hypotenuse or one of the shorter sides that you are looking for.

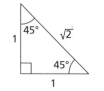

$$\tan 45 = 1 \qquad \sin 45 = \frac{1}{\sqrt{2}} = \frac{\sqrt{2}}{2}$$

$$\cos 45 = \frac{1}{\sqrt{2}} = \frac{\sqrt{2}}{2}$$

If you struggle to recall the standard trigonometric ratios, you can use triangles to find them. If you need to work with 45°, it is a right-angled isosceles triangle. Make the two equal sides 1. Pythagoras' theorem finds the third side as the square root of 2.

$$\sin 30 = \frac{1}{2} \qquad \cos 60 = \frac{1}{2}$$

so

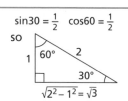

$$\sin 60 = \frac{\sqrt{3}}{2} \qquad \cos 30 = \frac{\sqrt{3}}{2}$$

$$\tan 30 = \frac{1}{\sqrt{3}} = \frac{\sqrt{3}}{3}$$

$$\tan 60 = \frac{\sqrt{3}}{1} = \sqrt{3}$$

To create this triangle, $\sin 30 = \frac{1}{2}$ so draw a right-angled triangle with 1 as the opposite and 2 as the hypotenuse. From this triangle you can read off all the trigonometric ratios for 30° and 60°.

$$\sin x = \frac{O}{H} \qquad \cos x = \frac{A}{H} \qquad \tan x = \frac{O}{A}$$

Make sure it is a right-angled triangle before using these ratios.

Label the sides of your triangle, starting with the hypotenuse. It can be easy to mix up the opposite and adjacent sides, so check you are getting it right.

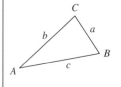

Sine rule $\dfrac{a}{\sin A} = \dfrac{b}{\sin B} = \dfrac{c}{\sin C}$

$$\frac{\sin A}{a} = \frac{\sin B}{b} = \frac{\sin C}{c}$$

Cosine rule

$$a^2 = b^2 + c^2 - 2bc\cos A$$

Area $= \frac{1}{2}ab\sin C$

Questions rarely give the sides the same letters as the formulae. You can rewrite the formulae with the letters you have or change the labels on the diagram.

The Quadratic Formula

$$x = \frac{-b \pm \sqrt{b^2 - 4ac}}{2a} \qquad \text{where } ax^2 + bx + c = 0$$

Make sure the quadratic is in the correct form before you identify the values of a, b and c.

Compound Measures

Speed $= \dfrac{\text{Distance}}{\text{Time}}$ Density $= \dfrac{\text{Mass}}{\text{Volume}}$ Pressure $= \dfrac{\text{Force}}{\text{Area}}$

When dealing with different measures, check the units carefully and convert if necessary before doing your calculations.

alternate angles a pair of equal angles formed when a pair (or more) of parallel lines are crossed by the same straight line (the transversal); the alternate angles are both on opposite sides of the transversal

alternate segment the 'other' segment; in a circle divided by a chord, the alternate segment lies on the other side of the chord

arc a curve forming part of the circumference of a circle

arithmetic sequence a sequence with a common first difference between consecutive terms

asymptote a line that a curve continually approaches but never touches

BIDMAS an acronym that helps you remember the order of operations: Brackets, Indices and roots, Division and Multiplication, Addition and Subtraction

binomial a polynomial with two parts; the sum or difference between two algebraic terms

box plot a representation of the distribution of a set of data, showing the median, quartiles and extremes of the data set

centre of enlargement the point from which the enlargement happens

centre of rotation the point around which a shape is rotated

chord a line joining two points on the circumference of a circle

class interval the width of a class or group, e.g. $0\,g < $ mass of spider $\leqslant 10\,g$

coefficient a constant multiplying an algebraic term that can be either a number or a letter

complete the square a specific form of a quadratic equation which can be used to solve the quadratic or find the turning point of the curve

compound interest interest that accrues from the initial deposit plus the interest added on at the end of each year

conditional probability the probability that an event will occur given that another event has already occurred

congruent exactly alike in shape and size

constant a number, either known or unknown, within an algebraic expression, that doesn't change

constant of proportionality the constant value of the ratio of two proportional quantities x and y; if $y \propto x$, then $y = kx$ (k is the constant of proportionality)

correlation the relationship between the numerical values of two variables, e.g. there is a negative correlation between the age and the value of cars

corresponding angles a pair of equal angles formed when a pair (or more) of parallel lines are crossed by the same straight line (the transversal); the corresponding angles are both on the same side of the transversal

cross-section/cross-sectional a shape/surface created when a cut is made through a mathematical object parallel to its 'base' / perpendicular to its central axis

cube number the product of three integers that are equal, e.g. $2^3 = 2 \times 2 \times 2 = 8$ so 8 is a cube number

Glossary

cumulative frequency the running total of frequencies calculated from a frequency table

depreciate/depreciation/depreciated a decrease in the value of something; generally expressed as a proportional reduction in value as a percentage

direct proportion two values or measurements may vary in direct proportion, i.e. if one doubles, then so does the other; this can be represented as a linear graph that passes through the origin with equation of the form $y = mx$ (the symbol $\propto$ means proportional)

empty set a set containing no objects (members)

enlargement a transformation of a plane figure or solid object that increases (or decreases) the size of the figure or object by a scale factor but leaves it the same shape; a scale factor of $-1 < f < 1$, $f \neq 0$ results in the shape getting smaller, while negative scale factors 'flip' the shape

equation a number sentence where one side is equal to the other

error interval the range of values that a rounded or truncated number could have had before the rounding or truncation, expressed using the inequality symbols

expression a statement that uses letters as well as numbers

exterior angle an angle outside a polygon, formed when a side is extended

extrapolation estimations that are beyond the range of given values, meaning they may be unreliable

factor a number that can be represented algebraically and can divide the whole to leave either an integer value or an expression without fractions

factorisation finding one or more factors of a given number or algebraic expression

Fibonacci sequence a number sequence found in nature; the sequence is formed by adding the previous two terms;
$u_{n+2} = u_n + u_{n+1}$

finite set a set which has an exact number of members

formula an equation that enables you to convert, or find a value, using other known values, e.g. Area = Length × Width

frequency density the vertical plot on a histogram; Frequency density = $\frac{\text{Frequency}}{\text{Class width}}$

frequency tree see **tree diagram**

function a relationship between a set of inputs and a set of outputs

geometric sequence a sequence with a common ratio

gradient the measure of the steepness of a slope: $\frac{\text{Vertical change}}{\text{Horizontal change}}$ or 'how many 'up' for every 1 space to the right', bearing in mind that −1 up is the same as 1 down

histogram a chart that is used to show continuous data using frequency density, i.e. using the area of the bars to represent the spread of data

identity similar to an equation but true for all values of the variable(s); the identity symbol is ≡, e.g. $2(x + 3) \equiv 2x + 6$

independent events two events are independent if the outcome of one event is not affected by the outcome of the other event, e.g. flipping a coin and rolling a dice

index (also known as **power** or **exponent**; plural: **indices**) the small digit to the top right of a number that tells you the number of times a number is multiplied by itself, e.g. 5^4 is 5 × 5 × 5 × 5; the index is 4

Glossary

inequality a statement showing two quantities that are not equal
Key symbols:
$x \neq y$ x and y are not equal
$x \leqslant y$ x is less than, or equal to, y
$x \geqslant y$ x is greater than, or equal to, y
$x < y$ x is less than y
$x > y$ x is greater than y

infinite set a set which continues forever, i.e. it has no end number

intercept the point where a line or graph crosses an axis

interest a percentage increase, generally associated with money invested in an account

interior angle an angle between the sides inside a polygon

interpolation estimations of values between known discrete data points; as the estimation is within given values, it is likely to be reliable

interquartile range the difference between the lower quartile and the upper quartile, often found using a cumulative frequency graph

intersection the point at which two or more lines cross

inverse (indirect) proportion two quantities vary in inverse proportion when, as one quantity increases, the other decreases

irrational number a number that cannot be written in the form $\frac{a}{b}$ where a and b are integers

iteration the repetition of a process in order to find an approximate solution; the result of one iteration is used as the starting point for the next

locus (plural: loci) the locus of a point is the path taken by the point following a rule or rules

lower bound the bottom limit of a rounded number

lower quartile the reading that is $\frac{1}{4}$ of the way up the cumulative frequency graph or a data set

mean an average value found by dividing the sum of a set of values by the number of values

median the middle item in an ordered sequence of items

midpoint the point that divides a line into two equal parts

modal class the class with the highest frequency in a grouped frequency table

mode the most frequently occurring value in a data set

multiplier the number by which another number is multiplied

mutually exclusive outcomes two or more outcomes that cannot happen at the same time, e.g. throwing a head and throwing a tail with the same flip of a coin are mutually exclusive events

percentage increase/decrease the change in the proportion or rate per 100 parts

perpendicular bisector a line drawn at right angles to the midpoint of a line

power (also known as **index** or **exponent**) the small digit to the top right of a number that tells you the number of times a number is multiplied by itself, e.g. 5^4 is $5 \times 5 \times 5 \times 5$

plane (figure) a figure or object that lies on a two-dimensional flat surface; two-dimensional geometry deals with plane figures

Glossary

polygon a two-dimensional shape that is constructed from a set of three or more straight edges

polynomial an algebraic expression that is the sum of algebraic terms that are of the form ax^b

prime factor a factor that is also a prime number

prime number a number with only two factors: itself and 1

probability tree see **tree diagram**

Pythagoras' theorem the theorem which states that the square of the hypotenuse of a right-angled triangle is equal to the sum of the squares of the other two sides; $a^2 + b^2 = c^2$

quadratic (equation) an equation containing unknowns (in a polynomial) with maximum power 2, e.g. $y = 2x^2 - 4x + 3$; quadratic equations can have 0, 1 or 2 solutions

quadratic (graph) the graph of a quadratic equation; the curve is smooth and symmetrical

range the spread of data; a single value equal to the difference between the greatest and the least values

ratio the ratio of A to B shows the relative amounts of two or more things and is written without units in its simplest form or in unitary form, e.g. $A : B$ is $5 : 3$ or $A : B$ is $1 : 0.6$

rational number a number that can be written in the form $\frac{a}{b}$ where a and b are integers

reciprocal the reciprocal of any number is 1 divided by the number (the effect of finding the reciprocal of a fraction is to turn it upside down), e.g. the reciprocal of $\frac{2}{3}$ is $\frac{3}{2}$

reflection a transformation of a shape to give a mirror image of the original

regular (polygon) a shape with all the edges of equal length, and all the angles are also equal

relative frequency
$$= \frac{\text{Frequency of a particular outcome}}{\text{Total number of trials}}$$

resultant (vector) the result of adding two or more vectors together

roots in a quadratic equation $ax^2 + bx + c = 0$, the roots are the solutions to the equation

rotation a geometrical transformation in which every point on a figure is turned through the same angle about a given point

scalar a quantity which has only magnitude

scale factor the ratio by which a length or other measurement is increased or decreased

scalene a triangle that has no equal sides or angles

sector a section of a circle between two radii and an arc

set a collection of objects (members)

similar a shape which is an enlargement of another with scale factor $\neq 1$

simple interest interest that accrues only from the initial deposit at the start of each year

simplify making something easier to understand, e.g. simplifying an algebraic expression by collecting like terms

simultaneous equations two or more equations that are true at the same time; on a graph, the intersection of two lines or curves

square number the product of two integers that are equal, e.g. $3^2 = 3 \times 3 = 9$ so 9 is a square number

square root the square root of a is the number that when multiplied by itself (squared) has a result of a

standard index form/standard form a shorthand way of writing very small or very large numbers; these are given in the form $a \times 10^n$, where a is a number between 1 and 10

stem and leaf diagram a semi-graphical diagram used for displaying data by splitting the values

subset a set within a set

substitution to exchange or replace, e.g. in a formula

supplementary angles angles that add up to 180°

surd a number written as a square root, e.g. $\sqrt{3}$; a surd is an exact number

tangent a straight line that touches a curve or the circumference of a circle at one point only

term in an expression, any of the quantities connected to each other by an addition or subtraction sign; in a sequence, one of the numbers in the sequence

translation a transformation in which all points of a plane figure are moved by the same amount and in the same direction; the movement can be described by a column vector

tree diagram a way of illustrating outcomes of an event, or combined events; branches are used to show the different outcomes. On a **probability tree**, each branch is labelled with the probability of the outcome for the single event and probabilities are calculated

by multiplying along the branches. On a **frequency tree**, the outcomes are shown on the branches and the frequency (number of) outcomes shown at the end of each branch; probabilities can be calculated using the relative frequencies

trial and improvement a method of solving an equation by making an educated guess and then refining it step-by-step to get a more accurate answer

triangular number an integer that can be represented in a triangle, e.g. the pins in ten-pin bowling: 1, 3, 6, 10, 15, ...

trigonometry the branch of mathematics that shows how to explain and calculate the relationships between the sides and angles of triangles by looking at the ratios of the sides

turning point in a quadratic curve, a turning point is the point where the curve has zero gradient; it could be a minimum or a maximum point

universal set contains all the objects being discussed

upper bound the top limit of a rounded number

upper quartile the reading that is $\frac{3}{4}$ of the way up a cumulative frequency graph or a data set

variable an algebraic term which can take different numerical values, generally in relation to a second variable as expressed in an equation

vector a quantity with both magnitude (size) and direction; it can describe a movement on the Cartesian plane using a column, e.g. $\begin{pmatrix} 3 \\ -2 \end{pmatrix}$ which means 3 right and −2 up (or 2 down)

Acknowledgements

The authors and publisher are grateful to the copyright holders for permission to use quoted materials and images.

Every effort has been made to trace copyright holders and obtain their permission for the use of copyright material. The authors and publisher will gladly receive information enabling them to rectify any error or omission in subsequent editions. All facts are correct at time of going to press.

Published by Collins
An imprint of HarperCollins*Publishers*
1 London Bridge Street
London SE1 9GF

ISBN: 978-0-00-822734-0

First published 2017
10 9 8 7 6 5 4 3 2 1
© HarperCollins*Publishers* Limited 2017

British Library Cataloguing in Publication Data.

A CIP record of this book is available from the British Library.

Cover and p1 © Cultura Creative (RF) / Alamy Stock Photo, © Shutterstock.com

Commissioning Editors: Katherine Wilkinson and Clare Souza
Authors: Rosie Benton and Jenny Hughes
Project Management: Richard Toms
Editorial: Richard Toms and David Mantovani
Proofreading: Alissa McWhinnie
Cover Design: Sarah Duxbury
Inside Concept Design: Paul Oates
Text Design and Layout: QBS Learning
Production: Lyndsey Rogers and Natalia Rebow
Printed and bound in China by RR Donnelley APS